Colorado Caves

Hidden Worlds Beneath the Peaks

WRITTEN BY
Richard J. Rhinehart

PHOTOGRAPHY BY
David Harris

WESTCLIFFE PUBLISHERS
www.westcliffepublishers.com

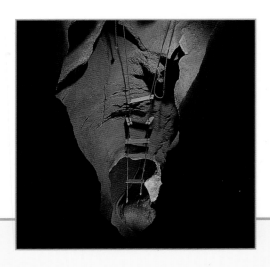

Contents

Acknowledgments 4

Foreword 7

Introduction 9

Colorado Caving Timeline: 1869–2000 16

Common Cave & Caving Terms 18

Chapter 1:
Great Explorations 20

Chapter 2:
The Legacy 36

Chapter 3:
Subterranean Science 52

Chapter 4:
Adventures in Surveying 70

Chapter 5:
Developing Commercial Attractions 82

Chapter 6:
Cave Conservation 96

Chapter 7:
Photographing in the Darkness 104

Chapter 8:
Visiting Underground Worlds 112

Afterword 123

Appendix A:
Colorado's Commercial Caves 126

Appendix B:
Caving Organizations and Websites 128

Appendix C:
Related Reading and Videos 130

Index 132

ABOVE: Reception Hall rope ladder, Cave of the Winds.

OPPOSITE: Beaded helictites, Heaven's Gate, Breezeway Cave.

PAGE ONE: June Harris peers into the west entrance of Hubbard's Cave.

ISBN: 1-56579-381-1

Editor: Jenna Samelson
Designer: Pauline Brown
Production Manager: Craig Keyzer

Published by:
Westcliffe Publishers, Inc.
P.O. Box 1261
Englewood, Colorado 80150
www.westcliffepublishers.com

Printed in China through:
World Print Ltd.

**Library of Congress
Cataloging-in-Publication Data:**
Rhinehart, Richard J., 1957-
    Colorado caves : hidden worlds
beneath the peaks / text by Richard J.
Rhinehart ; photography by David
Harris.
        p. cm.
    ISBN 1-56579-381-1
    1. Caves--Colorado. I. Harris,
David, 1961- II. Title.
GB605.C6 R45 2001
551.44'7'09788--dc21
                        2001017586

Acknowledgments

For 27 years, I have been pleased to consider myself a member of the Colorado caving community. Through times both good and bad, it is a privilege to have the opportunity to explore, study, and enjoy the many caves of Colorado with members of the National Speleological Society.

Over the years, my caving travels in Colorado and in other western states have been with dozens of fine cavers, and some of them are mentioned in this work. For those of you whom I have not mentioned, your efforts and friendship are still appreciated and cherished. Particular thanks must be given to the owners and managers of Colorado's two commercial caves, Cave of the Winds' Grant Carey and, at Glenwood Caverns, Steve and Jeanne Beckley and Phil Kriz. Colorado cavers are fortunate to have open access to such interesting places.

For the production of this work, I am indebted to numerous individuals. At the top of this list is Linda Doyle at Westcliffe Publishers, who had the idea for a Colorado cave book in the first place and contacted me to help her. My editor at Westcliffe, Jenna Samelson, helped me create a work greater than I might have believed possible when this project first began. Thanks also to John Fielder.

In researching the long history of Colorado caves, I must thank numerous individuals for their assistance and support. Among these are Donald G. Davis, Ken Kreager, Larry Fish, and Dr. Fred Luiszer. Thanks also to those who assisted in the selection of historic photographs: the Frontier Historical Society in Glenwood Springs, the Colorado Springs Pioneers Museum, the Cave of the Winds Collection, the Denver Public Library Western History Collection, and the Colorado Historical Society in Denver. I also wish to thank Carolyn Cronk, Dan Ridgeway, John Streich, and Dr. Norman Pace for use of their personal images of Colorado caving history. I give special thanks to Norm for writing this book's Foreword.

In addition, thanks to Dr. Fred Luiszer and Paul Burger, who provided graphic assistance for this work.

Finally, thanks to David Harris for his exceptional efforts in documenting the wonderful beauty and intrigue of the caves of the Columbine State. Not since the days of William Henry Jackson have we seen such a talented cave photographer.

—RICHARD RHINEHART
Greenwood Village, Colo.

I am very grateful for the kind and patient assistance of numerous Colorado cavers who helped make the photography in this book possible. In fact, it is entirely fair to say that without the help I received from these generous people, these photographs would not have happened at all. No one should cave alone and I would not have wanted it that way. I am glad to have shared the experience with my friends.

I would like to begin by thanking the following people for their generous gift of time and talent: Evan Anderson for his wonderful good nature in Glenwood Caverns and for sharing his expert knowledge of many delicate Colorado caves; Hazel Barton for her ability to stay *absolutely* motionless for much too long, her constant good humor, and her willingness to go just about anywhere at any time; Alan Williams and Dan Sadler for an incredible trip through Groaning Cave; Christy Harrison and Carl Bern for their unwavering devotion to doing the same shot again and again in Glenwood Caverns; L. P. Lawrence for helping in Breezeway and Cave of the Winds; Paul Burger; Steve Lester; and Richard Rhinehart for his open communication in locating many caves, and his assistance in finding historic photographs.

Much more than thanks are due to the owners of the commercial caves in Colorado: Grant Carey of Cave of the Winds for carte blanche access to all his beautiful caves and for his kind support of the strange ways I have photographed in them; Steve and Jeanne Beckley and Phil Kriz of Glenwood Caverns and the historic Fairy Cave for their kind patronage of my work and for trusting me with no less than the key to their cave.

Special thanks go to Bill Allen. He has shared so many adventures with me that we could write several books about them. Bill is an excellent photographer in his own right and I am grateful to him for years of friendship in our epic journeys together.

Lastly, I must single out my wife, June, for her help and encouragement on many trips around Colorado searching for obscure caves. With her help alone, we succeeded in some risky attempts to photograph the more dangerous places we found in many Colorado caves.

—DAVID HARRIS
Woodland Park, Colo.

OPPOSITE: **Fog enshrouds southern Colorado's Williams Canyon, home to many caves.**

Foreword

We don't have to peer far into the past, less than 150 years, to encompass most of the history of the human use of Colorado caves. As far as we know, Native Americans in the Rocky Mountains didn't make ceremonial use of caves or explore them extensively. Drawing on a general fascination that lures people to caves, European settlers launched the commercial development of Colorado caves. Some early commercial efforts function as popular tourist attractions today. Beyond the scope of the public eye, generations of cave explorers have ventured into Colorado's underground realms. Since the mid-1950s and the establishment of a Colorado caving community under the umbrella of the National Speleological Society, miles of virgin cave have been discovered and mapped, and the exploration goes on. The state is home to internationally noteworthy caves—some of the most beautiful and toughest anywhere.

Colorado Caves supersedes, by a generation, Lloyd Parris' *Caves of Colorado*. Published in the early 1970s, Parris' book provided a historical compilation and descriptive catalog of the state's caves. The impending publication of the book spurred a crisis in the close-knit community of Colorado cavers. Some argued it would identify targets for insensitive visitors, resulting in vandalism of fragile caves. Others welcomed public recognition; they believed that in the long run, Parris' message of conservation and introduction to organized caving outweighed the threat to the caves. This threat was and remains serious: Once-beautiful caves have been destroyed over the centuries. Unlike the scarring of a mountainside, which can heal with time and weathering, damage to cave walls and formations is permanent.

The caving community's furor was useful; Parris disarmed much of the criticism of his book by omitting specific locations of caves. Today, despite the controversy, you would be hard-pressed to find a Colorado caver without a copy of *Caves of Colorado*. The book drew some people to caves, but so did the population expansion of the state and the increased mobility wrought by new highways. Much of the exciting history of Colorado's caves has occurred since the publication of *Caves of Colorado*. Nonetheless, perhaps as a result of the conflict, there's been a dearth of public information on Colorado caves.

Richard Rhinehart and David Harris, longtime contributors to Colorado caving, now rectify this lack of information. In addition to an updated history and view of the caves, the authors invoke a paean to Colorado cavers. Unusual people are drawn to caving, sacrificing time, income, and usually some skin to explore, map, and conserve the state's caves. This book celebrates the efforts of cavers, and—thanks to the hard work of the author and photographer—you, too, can experience the beauty and excitement of Colorado's caves.

—NORMAN PACE, PH.D.
University of Colorado, Boulder, Colo.

ABOVE: 1950s Colorado Grotto trip, Wilson's Cave. PHOTO BY GLENN POLLARD, COURTESY JOHN STREICH COLORADO GROTTO COLLECTION.
OPPOSITE: White moonmilk coats a group of stalactites and columns in Glenwood Caverns' Paradise.

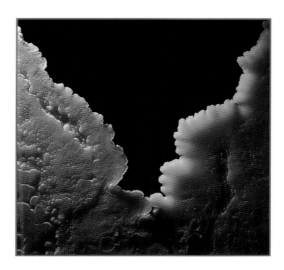

Introduction

It's a long way down. From the top, the entrance passage to Cave Creek Cavern is precipitously steep and foreboding. Loose dirt and rock nearly fill the passage, the result of years of debris tumbling down the mountainside into a convenient trap.

Actually, the upper entrance passage is not a natural cave at all, but an excavated tunnel opened in 1893 by miners more interested in precious metals than natural wonders. Its location southwest of Fairplay in the Mosquito Range places it in Colorado's mineral belt, where over the last century millions of dollars of gold, silver, molybdenum, and other precious minerals have been extracted. No doubt, their discovery of a natural cave, rather than a rich mineral vein, brought both surprise and disappointment.

Their new cave, however, raised the interest of Fairplay residents. During the next century, schoolchildren and adventurers spoke in awe of the cave's "big room." It was so large, they said, that the county courthouse could comfortably rest within.

It is September 1996. In the warm late-summer sun, I am hesitant to venture underground. Many out-of-state cavers who visit our Colorado caves remark how difficult it is to leave the beauty of the surface for the darkness of a cave. I'm inclined to agree, as the scenery of places such as the Mosquito Range, the White River Plateau, the San Juan Mountains, and Williams Canyon outside Manitou Springs is breathtaking in its own right.

Today, however, I am interested in seeing the big room of Cave Creek. I want to decide if it is larger than Manitou Grand Caverns' Grand Concert Hall, Groaning Cave's Shattered Hall, Glenwood Caverns' The Barn, or Breezeway Cave's Stone River. If it is, Cave Creek will earn the distinction of having the largest known underground chamber in Colorado.

Sliding feet first on my back down the incline, I can see how the passage became jammed with rock and dirt filtering down from the surface. We'll have to make sure cavers take care in manicuring the passage, sweeping excess dirt and rock into a 62-foot-deep blind shaft that the miners excavated at the foot of the incline.

Until our group of cavers reopened it during the summer of 1996, Cave Creek Cavern was considered "lost." Sometime in late 1975, the old timbered entrance collapsed, leaving no trace of the cave. Through careful study of the mountainside, using old maps and memories of cavers who had previously visited the cave, we were able to find in June the buried entrance with its twisted, broken timbers.

Rather than retimbering, our group of volunteers chose to "daylight" the tunnel, removing timbers, rock, and soil from the collapsed tunnel. This now provides direct access to the steep incline dropping down into the cave. Unfortunately, our hillside excavation also allows winter snow and summer rain to fall directly into the entrance. Ultimately, this may

ABOVE: Paper-thin layers of calcite, or rimstone, form along the edge of a pool in Breezeway Cave.
OPPOSITE: Stephen Reames perches on a flowstone mound near Stone River to admire Breezeway Cave's draperies.

cause the mountain to eventually slump further, sealing the cave once again.

There's no worry of that, however, during this September trip. At the foot of the incline, I carefully make my way around the miner's shaft, and follow my companions into a three-foot-high crawlway.

What makes someone want to go underground? Is it the lure of the unknown, the excitement and drama that accompany a trip into a cavern? Is it simply curiosity, an inherent need to understand the landscape and the natural forces that created our planet? Perhaps it is the timelessness of caves that is the attraction. While wind, rain, and snow can change landscapes perceptibly on the surface, changes in caves may take hundreds or thousands of years.

Here in Cave Creek Cavern, only one group has preceded ours since we reopened the cave earlier this day. Yet, there are signs of trips and visitors from a quarter-century or more before ours, appearing as if they were placed yesterday. I recall that in 1975, just prior to the collapse, I had searched unsuccessfully for the cave with high-school friends. Though clearly the outside world—and I, for that matter—had changed considerably since that August afternoon, here in Cave Creek, time had seemingly stood still.

My companions lead me through a series of low-ceilinged crawlways, where I have to drop down to my belly at times. At the end of the line of cavers, I can only catch occasional comments from my companions: generally talk about friends and family. With me are members of Conifer's Mayne family, a hardworking group of young cavers who provided much-needed assistance in our reopening of Cave Creek.

Leading our way is Karla Mayne, who earlier had been the first to descend into the cave following our excavation of the narrow entrance slide. Her father, Karl, had tied a rope around her waist and instructed her to shout if she wanted us to pull her out. Bravely, Karla slid down into the unknown, her father poised at the top of the slide to provide assistance at the first sign of trouble. Fortunately, she was able to kick her way through the debris plugging the slide and came upon the miner's shaft at the foot of the passage. This landmark confirmed our success in reopening Cave Creek. With a little bit of work, she cleared out more of the debris from the slide to allow the rest of us to descend in two separate groups.

Our group moves rapidly through the cave. The passage generally requires a range of movements, from crawling on

Cavers prepare to rig a large chute to aid in dirt removal from Narrows Cave.
PHOTO BY CAROLYN ENGLUND CRONK.

hands and knees to maneuvering through a low stoopway. We pause for a moment at a pleasant display of stalactites and flowstone, and then continue on our journey deeper into the cave. From a side passage, a small trickle of water enters. I recall from the 1961 Colorado Grotto map of the cave that the big room is only a few minutes ahead.

Colorado Caving

Cave exploration and study in Colorado in the latter 20th century grew remarkably, in part because of the Colorado Grotto's influence. The state's oldest chapter of the National Speleological Society, the Denver-based caving club helped bring order and direction to the exploration, survey, and scientific study of Colorado's caves. Chartered in November 1951, the Colorado Grotto is the state's largest club, with more than 150 active members. Through the decades, the Colorado Grotto has been involved with nearly every major discovery in the state, as well as exploratory and scientific endeavors in caves from Montana to New Mexico to old Mexico and overseas. Though other society-chartered chapters exist in towns like Glenwood Springs, Fort Collins, and Colorado Springs, the Colorado Grotto is still the state's most active group.

Like most caving groups nationwide, cavers in Colorado can be very secretive about their discoveries and the caves they explore and study. While new members are usually welcomed, cavers generally do not advertise their interest in

OPPOSITE: June Harris ventures into the Two-Level Room of Fulford Cave.

the underground, nor do they actively seek publicity in the popular media. This tendency reflects the delicate nature of caves and the potential for overuse and abuse by unknowing or uncaring visitors.

Unlike cave-rich states such as West Virginia, Missouri, and Tennessee, Colorado is home to relatively few caves. There are only 10 caves with more than a mile of surveyed passage; the great majority of the state's known caves have less than 1,000 feet of passage. Hundreds have less than 100 feet.

The public's interest in the caves of Colorado naturally is greater regarding the state's longer and more complex caves. While Cave of the Winds and Glenwood Caverns are commercially operated as visitor attractions and receive additional protection, the U.S. Forest Service and the Bureau of Land Management oversee the majority of the others. In the public's interest, these government agencies must manage resources on their lands to the best of their abilities. Public-owned caves deemed to be "significant" are provided special protection from abuse through the 1988 National Cave Protection Act. Violations of this federal law are punishable through fines and jail terms.

Cave scientists consider most Colorado caves to be low-impact. That is, it takes only a small amount of intentional or unintentional human impact to bring lasting damage to the resource. Damage can range from breaking delicate speleothems—stalactites, stalagmites, columns, helictites, and other such formations—to tracking mud onto pristine floors or destroying underground life.

A Sad Relic

Near the city of Glenwood Springs, along the northern wall of scenic Glenwood Canyon, is a sad example of a once-beautiful cave. Following a century of abuse and neglect, Cave of the Clouds overlooking Interstate 70 seldom sees visits from cavers.

Once a commercial attraction, Clouds is a shadow of its former self. Visitors to this once-wondrous cave have chipped and smashed nearly every stalactite and stalagmite that earlier delighted paying visitors. Spray-painted names cover the walls. Dust kicked up from countless visitors covers the remaining formations with a uniform gray color. Even if the owning Bureau of Land Management decided to build a gate on the entrance and restrict access to a select few, the cave will never again regain its former glory. It stands alone and

abandoned, a reminder to all who venture underground about how humankind can soil some of earth's most beautiful areas.

A World of Extremes

Following the intersection with the side tunnel channeling the small stream, our passage leads downhill at an increasing angle. We move with care across the rough, water-washed floor, avoiding the stream whenever possible. While in late August the stream is but a trickle, I suspect in late spring and early summer it is a torrent of frigid water, fed by melting snowfields on the mountains above.

Surprisingly, not many caves in Colorado contain streams or lakes. Most are dusty dry or filled with boot-sucking mud. The sound of water flowing underground is melodic, a pleasing change from the exceptional quiet of most caves. The beating of my own heart has surprised me more than once when waiting alone for the return of companions exploring side passageways.

The complete and total darkness of caves can be intoxicating. Although some commercial cave tour guides claim the darkness is a certain percentage darker than the darkest night, it is technically much more than that. Caves are darker than any overcast night or even deepest outer space, for there are no distant stars or planets to bring forth light. The darkness underground is utterly black without the tiniest amount of light, no matter how long you wait for your eyes to adjust.

Caves at Colorado's high altitude are also cold. They remain a constant temperature throughout the year—an average of the highest and lowest temperatures of the outside world. The air is the same temperature as the rock, meaning that a cave at 38 degrees, like many Colorado caves, can bring about hypothermia and even death if the visitor is not properly dressed. Just as one would not consider spending a night outside in the mountains without warm clothing, an inadequately dressed visitor to a cave will become very cold in a short amount of time.

One reason for the coldness is the constant humidity in caves; in many, the humidity hovers around 100 percent. This is much moister than the outside world where Coloradans enjoy days with 20 percent or lower humidity. With such high humidity underground, the cold feels even more so. Yet, it is the constant humidity that allows continued growth of stalactites, stalagmites, and other speleothems.

OPPOSITE: Bill Allen lights the ascending passage above the entrance to Cave of the Clouds.

Carl Bern studies a group of draperies formed on the lip of an overhanging breakdown block in Glenwood Caverns.

The Big Room

It is chilly following the trickle in Cave Creek Cavern. My coveralls and gloves are wet from the moist bedrock. We continue downward, our voices echoing curiously from some large chamber just ahead. I hear some shouts of exhilaration from my companions—undeniably, we have reached the cave's big room.

To my delight, it is a big room. We emerge from our descending passage onto a modest rock balcony overlooking the mighty chamber. Stretching beyond the limits of our electric lights, the room towers over our heads as well. From somewhere below us, I hear the splashing of a waterfall. The Maynes are already finding routes down off the balcony to the rocky, uneven floor of this great hall.

Taking in the view, I decide this room is larger than the others. It is certainly the largest natural underground chamber any caver has seen in Colorado. Surveyors who measure and chart the room weeks later resort to helium balloons with measured cords to determine a height of more than 90 feet. With a taped length of more than 240 feet and a width of greater than 80 feet at its largest point, this room is one for the record books. Officially, it is named the Colorado Room in honor of the room's immense size and volume.

I investigate the stream, which disappears noisily into large rocks along a wall. Others examine large wooden poles that have fallen from scaffolding dating from the early 1900s. Most likely, miners built this now-shaky scaffolding in an attempt to remove attractive stalactites from a major ledge far above the room's floor.

Hearing distant voices, I look up to see two cavers who have made their way up the far wall of the room. They look like distant stars in the darkness, making their way up a steep, rocky scree slope toward the Crystal Cliffs, a drapery barely visible in my bright electric light. Soon, they turn back as the hill becomes too steep and exposed for their ability.

Gathering together again, we examine curious artifacts on a flat rock table near the entrance of a small side passage. The materials seem to be trash from the 1970s, abandoned thoughtlessly by a previous generation. We take a few pieces of trash into our cave packs and leave the rest for a later clean-up effort.

Someone glances at a watch: It's time to head out. The rest of our party awaits our safe return at the entrance. We shoulder our packs and take our places in line, Karla leading the way.

Climbing to the balcony, I pause for a moment to take in one last view of the mighty Colorado Room. My electric light cannot begin to penetrate the darkness that envelops the now-quiet chamber behind me. It is a moment to remember, to cherish in the months and years to come.

Our efforts to reopen a forgotten cave and bring it back to the known world have been successful. Soon, Colorado cavers will understand this cave as well as any other in the state. Yet, as I turn and begin the long trip back to the sun-drenched surface, I feel some guilt.

Cave Creek Cavern and its magnificent Colorado Room were fine without human interference for a quarter of a century. The cave's collapsed entrance provided it with security that cavers and even the Forest Service cannot. But despite the cave's protection, its big room shows it is not without damage. Broken stubs of stalactites and stalagmites are common, as are candle-smoked signatures on the walls. While some damage is probably from the mining era, it is clear that 25 years of isolation cannot bring any healing.

As stewards for the natural features of this state both above and below ground, we must make the effort to see that there are no more victims of negative human impact like Cave of the Clouds in Colorado's future. Each cave, no matter how unimportant or insignificant it appears, must be fiercely protected. For unlike the forests of trees that cover the Rockies, or the air and water that flow freely across the surface, the underground world cannot be replenished, restored, or rebuilt within the lifetime of all humankind. Our actions and decisions in caves like Cave Creek or Cave of the Clouds will resonate for generations.

—RICHARD RHINEHART
Greenwood Village, Colo.

A trail of light from Bill Allen's headlamp marks his path through Cave Creek Cavern's immense Colorado Room.

Colorado Caving timeline: 1869–2000

1869 Cave of the Winds discovered by Arthur Love, a homesteader in Williams Canyon

1875 Mammoth (Huccacove) Cave discovered by quarrymen and surveyed by Colorado College students; W. S. Case opens it to public tours as first commercial cave in Colorado

1880 Brothers John and George Pickett rediscover inner Cave of the Winds; resulting publicity by the Reverend Roselle T. Cross and others encourages Charles Cross and the Boynton brothers to open the cave as an attraction, but high entry fees fail to attract visitors so the cave is closed

1881 George Snider further explores Cave of the Winds and reopens the cave with Charles Rinehart to public tours

1885 Manitou Grand Caverns opens public tours

First cave protection law in the United States passed by Colorado State Legislature

Williams Canyon and Cave of the Winds entrance building, circa 1885. PHOTO BY WILLIAM HENRY JACKSON, COURTESY COLORADO HISTORICAL SOCIETY.

1887 M. S. Yarwood opens Alexander's Cave (later Cave of the Clouds) in Glenwood Canyon to the public

1892 E. Maxfield, R. C. Koontz, and other miners discover Fulford Cave on Brush Creek; informal tours begin the following spring

Colorado's first known underground wedding takes place in Manitou Grand Caverns' Fairy Bridal Chamber

1893 Walter Devereux opens Vapor Cave in Glenwood Springs to the public

William Henry Hubbard, Charles Hubbard, and Griffith Jones discover Hubbard's Cave in Glenwood Canyon during a hunting and prospecting expedition

1895 The Cave of the Fairies (now Glenwood Caverns) discovered above the city of Glenwood Springs

1896 Charles W. Darrow opens The Cave of the Fairies to public tours, and nearby Cave of the Clouds closes to tours

1897 Electric lights installed in Cave of the Fairies, the second cave in the United States to offer electrically lit tours

Exclamation Point at The Cave of the Fairies, circa 1905. PHOTO COURTESY SCHUTTE COLLECTION, FRONTIER HISTORICAL SOCIETY.

1907 Cave of the Winds installs electric lights for its tour route

1911 Continued discoveries in Centipede Cave lead to the opening of the cave to the public as Manitou Cave; promoters D. H. Rupp, J. F. Sandford, and R. D. Weir take advantage of local publicity and a more favorable location to challenge the better-known Cave of the Winds

1921 First Colorado Mountain Club expedition to Spanish Cave discovers but does not descend the deep interior shaft now known as The Jug

1929 Ben Snider and Guy Boyd excavate the connecting Windy Passage between Middle Cave and Manitou Grand Caverns; Snider also pushes through the narrow Rat Hole in Cave of the Winds and finds a low, connecting passage to Middle Cave

1932 Carl Blaurock leads Colorado Mountain Club expedition to Spanish Cave and explores much of the known cave, including The Jug

1942 Cave of the Winds and Manitou Grand Caverns considered as air raid shelters; the caves are listed as shelters from nuclear attack into the 1960s

1951 The National Speleological Society charters the Colorado Grotto, the only chapter between the Mississippi River and the West Coast

A tight squeeze in the former Cave of the Fairies, now Glenwood Caverns, circa 1955. PHOTO BY JOHN STREICH.

1958 Colorado Grotto members discover extensive new passages in Spring Cave, including inner stream galleries

1959 Donald G. Davis and the Southern Colorado Grotto begin exploration of the caves of Marble Mountain, including Spanish Cave; several new caves are discovered

Cave of the Winds advertising pamphlet, circa 1935. COURTESY RICHARD RHINEHART COLLECTION.

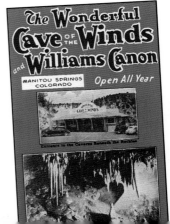

1960   Jam Crack discovery in Fairy Cave (formerly The Cave of the Fairies) by Pete Prebble leads Colorado Grotto members to The Barn, King's Row, and Paradise

1961   Colorado Grotto members Pete Prebble, Robert O'Connell, and Robert Wilber purchase Fairy Cave with intent to commercialize the cave

1968   Colorado School of Mines Student Grotto members John Pollack and Paul Westbrook find and explore the White River Plateau's Groaning Cave, soon to be known as Colorado's longest cave

1969   John Pollack and Steve Lewis discover Fixin'-to-Die and Wednesday Afternoon caves near Groaning Cave

Donald G. Davis, Lloyd E. Parris, and Robert Willis discover Premonition Cave on the White River Plateau; many other new caves are found as exploration and survey continue in Groaning, Fixin'-to-Die, and Premonition caves

1972   Landmark 1885 Colorado Cave Protection Act repealed by state legislature in "house cleaning" effort

1975   Colorado cavers join those from Utah, Wyoming, and South Dakota to form the National Speleological Society's Rocky Mountain Region

Al Collier and Tom Taylor successfully pass Spring Cave's Sump One; additional dives during the next three years extend the explored stream passage to Sump Four

Diving in Twenty Pound Tick cave on the White River Plateau, July 1979. PHOTO BY NORM PACE.

1980   Colorado Grotto member Bruce Unger, the only caver to perish in a cave, drowns in Goose Creek Cave near Deckers

1982   With approval from Cave of the Winds manager Grant Carey, Colorado cavers begin work in the cave on a variety of projects including exploration, surveying, conservation, and scientific study; cavers also explore other caves in the Williams Canyon area

Helictites, Breezeway Cave.

1984   Cavers Dave Allured, Richard Rhinehart, Fred Luiszer, Larry Fish, and others discover the remarkable Silent Splendor passage in Cave of the Winds

Rocky Mountain Caving quarterly journal founded by Colorado's National Speleological Society chapters

1986   Paleontological digs by the Carnegie Museum and the Denver Museum of Natural History (now Denver Museum of Nature & Science) begin in South Park's Porcupine Cave

Colorado cavers are wildly successful in their excavations of New Mexico's Lechuguilla Cave; their dig eventually leads to more than 100 miles of cave

1988   Cyndi Mosch, Rich Wolfert, and Tom Shirrell discover Hourglass Cave and the skeleton of an explorer who died there 8,000 years ago

1989   Colorado cavers form the Williams Canyon Project of the National Speleological Society to explore, survey, and study the caves of Williams Canyon

1990   Williams Canyon Project cavers discover the Mammoth Extension to Huccacove Cave, the cave's first significant addition since its discovery in 1875

1993   Colorado Grotto members Al Hinman, Fred Luiszer, and Richard Rhinehart discover Breezeway Cave in Williams Canyon; exploration and survey continue into 1994

1994   Jon Barker and Charles Lindsey discover the Deepwater Cave extension to Manitou Cave

1995   The "Lost Tour," a wild caving tour, begins at Manitou Cave under the direction of caver Marc Hament

1996   Led by Jon Barker and Charles Lindsey, Williams Canyon Project cavers discover significant passageways and rooms in Narrows Cave after years of excavation

The Leavenworth section of Narrows Cave. PHOTO BY CAROLYN ENGLUND CRONK.

1998   Cavers discover many new rooms and passages in Glenwood Caverns from the fall of 1998 through the summer of 1999, including Beginner's Luck, Polar Bar, and Discovery Glenn

1999   Steve and Jeanne Beckley and Phil Kriz open the former Fairy Cave as Glenwood Caverns; the tour includes lower-level passageways reached through a new entrance tunnel

2000   The discovery of extensive passageways in Groaning Cave brings its total surveyed length to 10 miles, reaffirming its status as Colorado's longest known cave

Evan Anderson exploring a labyrinth of passages deep within Glenwood Caverns.

Aragonite and calcite stalactites and stalagmites can be found in Glenwood Caverns' Paradise.

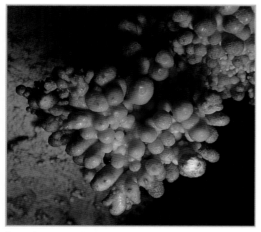

Drops of water on a cluster of calcite botryoids in the Grape Room of Hubbard's Cave.

Calcite "cave bacon" and stalactites in Cave of the Winds' Temple of Silence.

Common Cave & Caving terms

Aragonite: A calcium carbonate mineral occurring in a different crystalline form than calcite; in Colorado, commonly forms beaded helictites.

Ascenders: A variety of metallic mechanical devices used to climb ropes.

Blowhole: A natural hole in the ground from which air flows; can indicate a cave if barometric.

Botryoids: Calcium carbonate formations that resemble a bunch of grapes..

Breakdown: Rocks in a cave; can be large or small, alone or in piles.

Calcite: A calcium carbonate mineral forming most speleothems such as stalactites and stalagmites; also the primary mineral in limestone.

Carbonic acid: A weak acid formed from carbon dioxide and water that chemically dissolves limestone.

Cave, Cavern: A naturally occurring cavity in the surface of the earth large enough to admit humans and extend into total darkness.

Cave bacon: Colorful flowstone strips with parallel bands resembling bacon; found along cave roofs or walls.

Cave coral: A calcite speleothem that resembles popcorn or grapes; sometimes called botryoids.

Cave hunting: Seeking new, undiscovered caves on the ground's surface.

Cave pearls: A rare, spherical, calcite speleothem resembling pearls; formed through the action of water.

Cavelet: A small solution hole or other feature too tiny to admit humans or extend into total darkness; sometimes called a shelter cave.

Caver: An individual who visits the underground, including cave scientists (speleologists), photographers, explorers, conservationists, and surveyors; neophyte visitors may be called "spelunkers," an antiquated 1950s term for cavers.

Chimney: A natural crevice or passage in which cavers must climb either horizontally or vertically with their feet or hands on one wall and backs on the opposite wall; these features can be narrow or wide, sometimes with great depth.

Column: A carbonate speleothem formed when a stalactite and stalagmite grow together, often connecting the ceiling of a cave with the floor.

Commercial cave: A cave open for public tours as a paying business; may include developed trails and electric lighting.

Dig: A place in a cave filled with dirt, mud, or rock that can be excavated to reveal additional passage.

Dogtooth spar: A crystalline feature resembling teeth.

Dolomite: A magnesium carbonate sedimentary rock less soluble than limestone.

Dome: A cylindrical shaft in a cave ceiling; sometimes accompanied by a pit below.

Draperies: A calcite speleothem that looks like curtains, often with folds; one variety is called cave bacon, as it resembles strips of bacon.

Dripstone: A speleothem formed through the dripping of water; includes draperies, stalactites, and stalagmites.

Flowstone: A calcite speleothem formed by films of moving water that can cover walls, rocks, and floors of caves.

Grotto: A chapter of the 11,000-member National Speleological Society; also, a small shelter cave.

Guano: The solid waste from bats.

Gypsum: A hydrous calcium sulfate sedimentary rock more soluble than limestone; can be found as crusts or deposits in caves.

Gypsum flower: A gypsum speleothem growing in flower-like forms; variants include "hair," crystals, and needles.

Helictite: An aragonite or calcite speleothem, erratically developed in defiance of gravity; can be in many forms, including beaded, root, needle, and quill.

Joint: A fracture in the rock that does not include displacement but is favorable to cave development; often associated with faulting.

Karst: A landscape containing caves and other subsurface features such as sinkholes, springs, and disappearing streams; often, the limestone or other soluble rock is directly exposed on the surface as pavement (a type of exposed bedrock), pillars, and cliffs.

Lead: A cave passage that might be extendible through squeezing, climbing, digging, or simply investigating.

Limestone: A calcium carbonate rock of marine origin composed of mud, algae, and other shallow sea remains; this rock can be more or less soluble, depending on the amount of calcite present.

Moonmilk: A curious mineral consisting of hydromagnesite and water. Appearance generally ranges from plaster of Paris to cottage cheese; may be related to biological processes.

Pit: A natural shaft in the floor of a cave that may be vertical; sometimes beneath domes directly above.

Rappel: A method of descending a vertical pit or steep incline using rope and mechanical descenders; cable ladders and body rappels are seldom used by cavers today, though hand lines are used in less exposed or shorter pits.

Rimstone: A calcite depositional wall that holds, or has formerly held, water.

Sinkhole: A natural bowl- or funnel-shaped depression in the ground that may lead to cave passage or drainage below; these features can sometimes be opened through digging.

Solution: The chemical process of acidic water dissolving limestone and other soluble rocks and minerals.

Speleology: The science of caves and karst.

Speleothem: A secondary mineral deposit usually of calcite, aragonite, or gypsum; also called formations.

Stalactite: A speleothem hanging from the ceiling of a cave or mine; forms through the slow dripping of mineral-rich water.

Stalagmite: A speleothem growing from the floor of a cave or mine; forms through the slow dripping of mineral-rich water.

Virgin cave: Cave passageway that has never been visited or explored by humans.

Water table: The top level of the water-saturated zone of the Earth's crust; caves formed below this level are called phreatic.

Wild cave: A cave that does not have any visitor improvements such as trails, stairways, and lights.

Rapidly flowing spring run-off carved this large piece of flowstone in Spring Cave.

Helictites in Elkhorn Chambers of Breezeway Cave.

Bill Allen studies pristine calcite stalactites near Holy Waters in Breezeway Cave.

Great Explorations

Deep below western Colorado's Iron Mountain, 30-year-old Evan Anderson patiently sketches the irregular cave passageway before him. His companions talk quietly every now and then, breaking into easy laughter.

With precise detail, Evan draws the cave walls, floor, and features, including any dripstone formations, fossils, or other geologic curiosities. His pace is leisurely, though with deliberate purpose, for plenty of unknown cave awaits exploration and survey ahead.

It is midsummer and while thunderheads build in the warm skies above the bustling city of Glenwood Springs, here in the depths of Glenwood Caverns, time takes on a new meaning. With only a faint breeze to guide the team through the darkened passages, Evan declares he is ready and the rest of the team moves a few feet to the next survey station.

"On point," the lead man announces.

"Tight," instructs the compass man, who holds the survey tape firm on his station.

"Tight," responds the lead man, holding the "dumb end" of the tape—the tape's beginning—on the protrusion he has selected as the new station.

Evan waits patiently, perched on a large rock in the low corridor. In his notes, he will record the length of the survey shot, then the azimuth (or direction) along with the inclination (the angle of the passage). While these numbers are determined, he begins sketching the passageway with his mechanical pencil, tentatively at first, but with bolder marks once he receives the compass information. With a ruler and plastic protractor, Evan will sketch the passage to scale, providing a realistic and accurate plan view of the passages the team is surveying.

Glenwood Caverns, the former Fairy Cave, is a privately owned commercial cave that opened in May 1999. From the gift shop and museum near the historic Hotel Colorado, buses shuttle visitors up and down the steep, switchbacking Transfer Trail to nearly the top of the mountain. Tens of thousands of visitors annually tour the half-mile of passageways open to the public. Electric lights expertly illuminate these grand corridors and chambers, bringing a sense of adventure to both young and old.

True adventure, however, lies just beyond the reach of the electric lights. Here, in the utter darkness that envelops the underground, more than two miles of passageway snake through Iron Mountain. Many of these corridors, crawlways, and chambers have only just been found in the last few years by cavers, that odd lot of underground enthusiasts who enjoy forcing their bodies through squeezes in solid rock measured in mere inches, or digging through ancient debris until uncovering an opening. Cavers are driven by the sense of exploration and wonder, of the need to know all there is about a particular cave or passage.

ABOVE: In Glenwood Caverns, Hazel Barton makes notes in a survey book while Dan Lins and June Harris measure the distance between survey stations.
OPPOSITE: Golden aspens glisten outside the Grizzly Creek Canyon entrance to Bair Cave on the White River Plateau.

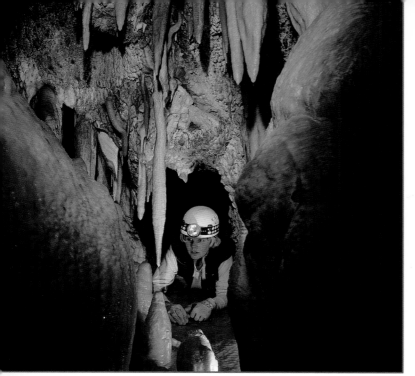

Christy Harrison enters a formation-filled passage in the Beginner's Luck section of Glenwood Caverns.

At Glenwood Caverns, a "survey-as-you-go" doctrine governs exploration. This is a recently developed philosophy, born in the 1970s and 1980s as the number of new cave discoveries dwindled and new, "virgin" passage became increasingly scarce. There are exceptions to this decline in the discovery of new caves. More than 100 miles of cave were discovered, explored, and surveyed in southern New Mexico's Lechuguilla Cave in the 12 years since its 1986 discovery by Colorado cavers. Even there, in corridors as large as highway tunnels and chambers larger than airplane hangars, "survey-as-you-go" dictates the exploration process.

Prior to today's philosophy, cave explorers delegated the task of surveying caves solely to the surveyors. These cavers delight in moving slowly through previously "scooped" cave passageways, documenting and measuring areas that are already known. In a single day, a good team can survey several hundred feet of passage. With such assistance, the true explorer never surveyed, or if he or she did, only when it was necessary to determine relationships between rooms and passages or separate caves.

Colorado's longest cave, Groaning Cave, was largely explored by only a handful of cavers. Presented with a complex labyrinth of crossing passages, these explorers literally ran through these unknown lands in the summer of 1969. These cavers found their way back out of the interconnecting passageways by leaving rock cairns, playing cards, and other temporary trail markers (string and marked arrows on the walls are impractical and forbidden). Survey teams of three to five followed unhurriedly behind, using the cairns as markers to find the main route into the cave.

Of course, such hurried exploration sometimes leads to the missing of key passageways that can later prove helpful. Groaning Cave's famous Conning Tower, for instance, stood for many years as a barrier to the cave's back regions. As the only known route leading to more than a mile of surveyed corridors and rooms, the Tower required all passing cavers to free-climb a steep, exposed rock crevice, then drop into a series of small pits—the last overhang requiring a roped descent—to reach the maze on the other side. Cavers had to carefully examine the rigged ropes on the drop to be certain of their integrity. On more than one occasion, damage to the rope from previous use was discovered.

Mindful that continuing rope problems could cause a serious accident, and interested in finding a nontechnical route to the back of Groaning, I spent an afternoon in July 1986 carefully examining the area on the Tower's near side. One of the first passageways I checked, a tight crawl leading directly from the beginning of the Tower ascent, proved to be the elusive bypass route. Early explorers, perhaps excited at the discovery of the cave's first and only rope drop, didn't check the obvious passage at its foot to see if it led anywhere interesting.

Accidental Discoveries

It's been said many times, "Caves are where you find them." Cavers sometimes spend hundreds of hours digging or squeezing or trying to force a passage to do what they want rather than allowing the cave to lead them. In theory, it should have been straightforward to connect Cave of the Winds' lower-level Breakdown Room with Manitou Grand Caverns' middle-level Grand Concert Hall. Surveys indicated that the two passageways were only 100 feet apart with an elevation difference of less than 50 feet. Such a connection would provide an alternate route to the Grand Caverns; the functioning route, the Windy Passage, was excavated in 1929 to provide access to the long-closed cave. In addition, our projected new route would allow for the development of a loop tour.

OPPOSITE: Stalagmites intrigue Alan Williams in Groaning Cave's CSU Passage.

In November 1983, cavers began digging in a mud-filled passage leading west from the Breakdown Room. According to the surveys and the cave's geologic fracture pattern, the passage should have intersected the Concert Hall. About halfway into our journey, we found that the rock ceiling of the horizontal passage had disappeared. In its place was sticky red clay, easily removed with a shovel and bucket. Knowing an ascent to the cave's middle level would need to be done sooner or later, digging upward seemed wise. We hoped this would reveal an open passage leading most of the remaining distance.

Unfortunately, the discovered passage did not cooperate. Though our dig did indeed break into a new chamber—the Whale's Belly—we never accomplished a connection to the Concert Hall. Instead, we scaled the Whale's Belly's high southern wall with rock-climbing bolts and associated technical gear to reach an unspoiled upper-level corridor. Rather than leading west toward the Concert Hall, the passage followed the dip of the limestone south, toward the commercial tour trail.

This new passage was clearly virgin, and Larry Fish and other cavers cautiously picked a route through the mud, following a single trail. For their efforts, they were rewarded with astounding beauty, a passage treasured to this day—Silent Splendor. The January 1984 discovery of Silent Splendor and its remarkable beaded helictites was not expected or predicted. Yet, by following the cave and allowing it to show us the way, we made one of the century's most exciting cave discoveries.

Not far from Cave of the Winds, I had the good fortune to participate in an even greater discovery a decade later. In a region of Williams Canyon that countless explorers had hiked and examined since the 1860s, our small group came upon an overlooked blowing hole. Small openings often leading to significant caves, blowing holes are particularly interesting if they are the result of barometric pressure and not just a chimney leading from a higher to a lower entrance. On that April 1993 afternoon, Al Hinman, Fred Luiszer, and I were searching the hillside for possible indications of a hidden cave. After only a few minutes, Al announced he had found a blowing hole next to a cliffside deer trail. Skeptical, Fred and I were surprised to feel the wind blowing onto our faces when we bent down to take a look.

During the next two months, our small group spent five Saturdays digging in the enlarged hole. After the mysterious breeze, we removed rock, dirt, and dusty packrat midden that

Stalactites, stalagmites, and frostwork formations detail the Silent Splendor area of Cave of the Winds.

jammed the cave. Passing beyond an unsteady area of large, loose boulders, Boulder caver Rene Singer led us into an immense, echoing chamber we called Cowboy Heaven. This airy room stretched beyond the combined brilliance of our electric lights. Cautiously we explored, taking care in where we walked so that future visitors could enjoy the chamber's wild nature.

The room led north to a pit, Hangman's Hole, which we carefully climbed. From its base, a fragile squeeze through stalactites and columns—the Jailbars—led to another towering canyon we called Happy Trails.

Our discovery overwhelmed us. By following the wind and believing in our dig, we found a major new cave. Within a year, Breezeway Cave was explored beyond a mile in length, revealing many marvelous sights, including beaded helictite displays that outshine the beauty of the famous Silent Splendor.

OPPOSITE: Delicate beaded helictites cling to a wall in Cave of the Winds' Silent Splendor.
NEXT PAGE: In Breezeway Cave, Harvey DuChene watches as Bill Allen uses a safety handline to descend into Hangman's Hole.

The Leadville Limestone of the White River Plateau holds numerous caves.

For cavers, the discovery and subsequent exploration of Breezeway Cave remain the crown jewels of our activities in the privately owned Williams Canyon. Previous and future discoveries in the canyon will invariably be compared to Breezeway and its many curiosities.

Summer's End Is Spring's Beginning

In the alpine limestone of the White River Plateau, I have spent countless weekends during the last quarter-century searching for another Groaning Cave. With 10 miles of surveyed passage, Groaning has been Colorado's longest known cave since August 1968, when cavers from the Colorado School of Mines investigated its entrance corridor.

Though my searches have revealed several interesting small caves, none possess the length, complexity, or beauty of Groaning. Searching for unknown caves is an acquired habit, particularly when there are known caves with possibilities for significant extension through excavation or careful pushing beyond tight squeezes. Called "cave hunting," this activity does not require a license or any particular training. Rather, it requires exceptional patience and the determination to return again and again to areas previously examined by others and declared finished.

In the rolling high-altitude meadows of the White River Plateau, many cavers have long sought the source of the underground stream that emerges more than 1,700 feet below into Spring Cave along the South Fork of the White River. Spring Cave has been visited for a century, though its huge inner passageways were only discovered in the last 40 years. Trained cave divers in the late 1970s and early 1980s ventured beyond four water-filled passages, called sumps, in search of the stream's elusive source. Yet, even in their deepest penetration, they found no indication of the stream's descent from a surface sinkhole or swallow.

Not having the training or the talent to follow water-filled passages underground, my group decided to approach the problem from another direction. If the divers were unable to follow the stream to its entry into the cave, perhaps we could find the location where the stream disappeared in its journey underground.

On a bright, sunny October day in 1980, Dave and Vi Allured, Steve Dunn, and I braved the early morning chill of the high plateau to examine its exposed limestone. On previous weekends, we had found several small fissures in the

Gary Ludi assists Tom Taylor in suiting up for a 1975 dive in Spring Cave's Sump One. PHOTO BY NORM PACE.

rock, though none extended for any distance or held any water. We were lucky, however, and came upon a small sinkhole in the meadow with an open hole at the bottom. Pleased at our good luck, we scampered inside and found ourselves in a spacious cavern leading steeply downward.

With due care, we climbed through a chaotic pile of boulders, some as large as automobiles, to find a way to the cave's lowest level. The passage grew in size, surprising each of us. Somehow, we expected it to end without warning. Just as we began to imagine finding an open passage descending to the back of Spring Cave, the cave abruptly ended in a grubby passage that evidently floods during the spring snowmelt.

Though Summer's End Cave held no running water, only a small, grimy pool at its lowest level, we believed it could very well be one of the sources for the Spring Cave stream. In September 1996, cavers with fluorescein dye turned the pool bright green—and a few weeks later, received a positive indication from special dye bugs placed in the Spring Cave stream of a hydrological connection.

Patience Has Its Virtues

Cavers willing to take the time to examine passageways more carefully than those who traveled before them can sometimes extend even well-known caves. Up until the early 1960s, the commercial Glenwood Caverns was a very popular wild cave open to those with a sense of curiosity. Although the cave operated as a commercial attraction for 15 years at the turn of the century, the entrance gate to the cave had long since disappeared. Adventurous visitors walked the former tourist trail and the tunnel leading to a Glenwood Canyon viewpoint.

Hazel Barton negotiates the Jam Crack squeeze, formerly the only way into the large, decorated rooms within Glenwood Caverns.

Denver cavers, however, had followed the elusive wind rather than the well-established commercial trail. By enlarging obscure holes through digging, hammering, and squeezing into tiny passages, Pete Prebble, George Moore, John Streich, and others discovered hundreds of feet of new passage.

Their efforts paid off generously after pushing beyond an obstacle called the Jam Crack. A vertical crevice with a reputation for tightness and length, the Crack led down 30 feet to a lower level where the wind whistled through a hole too small for the explorers. Hammering their way through this obstacle called Purgatory, the cavers found that the passage quickly expanded into huge passageways, now shown on the commercial tour. The Barn, with a towering pile of rock, seemed particularly colossal after the cramped passageways of

June Harris wades through the first water-filled passage beyond Jones Beach in Spring Cave.

the tortuous connection route between the upper and lower levels. King's Row—with its exquisite draperies, flowstone, stalactites, and stalagmites—leads deeper into the mountain, suggesting that the cave's ultimate destination is the water table below Glenwood Springs.

It is from King's Row that cavers nearly 40 years later made the new discoveries that Evan Anderson and his team surveyed in the summer of 1999. With the cave's reopening that spring, a new entrance tunnel was mined directly into The Barn. This tunnel provides easy access to lower-level passages that once were over two hours from the entrance.

In the Gypsum Halls, a low-ceilinged extension with extensive gypsum deposits, two cavers spent several weeks excavating an obscure crawl. In early July 1999, they finally broke into unknown open passage. Though Glenwood Caverns has an established policy regarding new discoveries, the excited cavers could not restrain themselves from exploring some of the new cave. Beyond some tight squeezes, the cave opened into great rooms and silent corridors, each with possible continuations. It was time to start surveying.

The Gypsum Halls breakthrough was one of many in Glenwood Caverns during 1998 and 1999. Cavers from

The Name Game

Each year, nearly 200,000 visitors tour the winding passageways of Cave of the Winds, one of the more successful privately owned caves in the country.

From Boston Avenue to the Crystal Palace, through the Old Curiosity Shop and on to the Temple of Silence, visitors hear stories of the geology and the discovery of these rooms and corridors. At the entrance to the Valley of Dreams, visitors examine a line of poetry along the wall, placed by the cave's management in gratitude to the individual who provided the Depression-era funding to open that passage. Yet, despite the many stories the tour guides tell along the way, hardly a word is uttered about the names of the rooms.

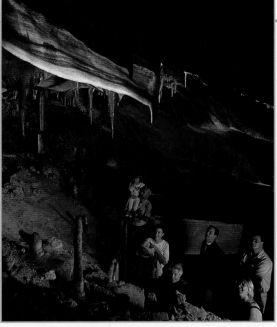

A tour group visits Cave of the Winds' Temple of Silence, named through a public contest in 1929.

In Cave of the Winds, historians have found that many rooms are named after familiar features from the outside world. Boston Avenue, for instance, is said to have received its name from tour groups who reported that it reminded them of the narrow streets of that great eastern city. The Temple of Silence found its name through a public contest in which thousands of entries were received in hopes of providing a name to the "new room" opened in 1929. Ordovician Avenue was named for the Ordovician-age Manitou Limestone in which the corridor developed.

Caves and passageways discovered and explored in recent times often bear more imaginative names. In lower Cave of the Winds is the Valley of the Shadows, a steep passage guarded by a huge, ominous boulder. Off of the Grand Concert Hall in Manitou Grand Caverns, a corridor was dubbed the Yukon Trail as it leads north to Iceland, a well-decorated chamber with pure white dripstone, and to Anchorage Pit, a slippery shaft dropping to a lower level.

In Groaning Cave, named for an unusual groaning sound created when wind blows through a small restriction, cavers were exploring new lands at the same time as astronauts' first walk on the moon. Therefore, some rooms, such as Tranquility and Serenity halls, share their names with lunar features. A nearby cave, reached by a long rappel on rope or by an exposed ledge traverse, was named by its explorers Feels-Like-I'm-Fixin'-to-Die Cave, now generally shortened to Fixin'-to-Die or even Fixin'.

Like the caves, passages also sometimes receive humorous names. In otherwise uninteresting Dilation Cave north of Manitou Grand Caverns, Rodent Denture Passage—with its many tiny mice teeth—leads from the Bearing Strait, home to some curious round balls of mud. Greasy Traverse in Spring Cave warns of a slippery slope; Electric Armpit Crawl in Manitou Cave recalls the squeeze's low height; and Udder Amazement in Groaning Cave describes some spectacular mammillary stalactites. At times, clever plays on words are selected. In a section of Glenwood Caverns discovered in 1998, cavers named a region with disco names after hearing the 1970s disco rock song "The Hustle" at a local fast-food restaurant during dinner. Future historians will no doubt puzzle over names like Paragon Disco, Copacabana, and a tight passage known as the Huss Hole.

To encourage creativity, cavers often select themes for cave names or regions of caves. In Breezeway Cave in eastern Colorado, the cave's initial explorers imagined a western theme. Rooms and features were dubbed Cowboy Heaven, the Fence Line, Happy Trails, High Plains, and Box Canyon. Explorers in nearby Narrows Cave named sections after famous prisons and the rock band The Grateful Dead.

During recent years, caves on the White River Plateau have been given unusual names. Knowledgeable cavers can lead you to Ain't Much Cave, Twenty Pound Tick, Insignificave, Animal Dung Cave, and even Green Octopus Cave. Usually, though not always, cavers will stay away from descriptive names that may provide a clue to the cave's location. Fanciful names like Grieving Widow's, Blue Butt, Sinking River, and Spectre caves are the result.

One of the benefits of being the initial explorers of a cave or a passage is the opportunity to name it and its significant features. Of course, the sketcher in a survey crew is the ultimate authority, as that person has the right to overrule names suggested by colleagues.

Often, with minor discoveries, names are applied to everything, including stalactites and stalagmites. With larger discoveries, cavers will usually give up and apply only general names to areas, such as "The D Survey" or the "New Horizons" section. For a giant cave like New Mexico's Lechuguilla Cave, cavers have created a gazetteer of place-names for rooms and features within the 100-mile-plus cave. Otherwise, no one could keep track of the meanings of the cave's place-names or the hopes, dreams, joys, and frustrations behind them.

Mysteries of the Dark Waters

Spring Cave houses a stream with an unknown connecting passage to its source.

One of Colorado's great underground mysteries is the overall extent of Spring Cave, the state's longest and most significant stream cave. Popular with amateurs, Spring Cave is reached by an easy Forest Service trail that climbs the lush wall of the White River's South Fork. Most visitors explore only the cave's descending entrance series, which leads from the double entrance past a sturdy wooden ladder to a lower stream level. Thunder Road is a noisy underground stream that splashes and gurgles its way through unknown passages to its reemergence along the White River. Just upstream of Thunder Road rests a quiet emerald lake, mistakenly thought by many to be the stream's source. From here, most visitors return to the surface, their sense of adventure and curiosity satisfied.

Knowledgeable cavers, however, follow a series of obscure passageways much deeper into the mountain, reaching the stream far beyond the tranquil pool. At sandy Jones Beach, the Spring Cave adventure really begins.

Filling the passage from wall to wall, the stream here is much less friendly than at Thunder Road. Explorers require full wetsuits to continue, and the willingness to swim in chilly 40-degree water. From Jones Beach, the cave alternately grows and diminishes in size, at times towering impressively 50 or 60 feet above the floor, at other times dropping to low, noisy passageways. Steve's Duck is perhaps the most fearsome of these low-ceilinged passages, for in late summer, complete immersion is usually demanded to continue.

The end of many Spring Cave journeys is Sump One, located at the back of the grand stream corridor. Here, the ceiling lowers down to the deep green water, blocking any possible continuation without a diving mask, pressurized dive tanks, weight belts, and flippers. A cobbled sandbar near the sump's entrance provides a convenient spot for staging technical dives to the cave's deepest passages and for getting out of the water after the dives.

For decades, cavers theorized that Spring Cave's stream entered from somewhere near Cow Lake on the western White River Plateau. Because the plateau lies more than 1,700 feet in elevation above the cave, if a connection could ever be established, Spring Cave would be the deepest known cave in the country. In 30-plus years of searching, no disappearing stream or great shaft was found. Other than a few shallow sinkholes and minor caves, the western White River Plateau seemed to hold little possibility for an underground connection.

True to its name, the Emerald Pool in Spring Cave glows like a gem where the Thunder Road stream flows from under the passage walls.

Since November 1962, when the Colorado School of Mines Student Grotto braved early winter snows for Spring Cave's first comprehensive survey and diving expedition, experienced cave divers have been trying to find this elusive passage. In the process, divers have discovered thousands of feet of new passage that include five sumps, subterranean waterfalls and lakes, a vast above-water gallery, and a huge underwater gallery filled with gigantic boulders.

Spring Cave has yet to be fully explored because of the dangers of cave diving, though every few years a new generation of divers gives it a try. The risks are real, even for trained and experienced cave divers. The blinding silt that obscures the water in the cave's sumps can be particularly frightening.

In September 1988, Marty Smith and Dwayne Matias got separated from the rest of their diving party in Sump Three because heavy silt had reduced underwater visibility to a few inches. Without a set dive line to mark the way through the flooded passage, it took cool heads and careful exploration of the chocolate-colored water to find the correct route through the narrow, confusing sump. Their companions, having waited more than 30 minutes for the divers to emerge from a sump that normally took only a few minutes to swim, had given the missing divers only a 50 percent chance of making it out alive. Thankfully, the divers emerged to cheers and the group headed for the cave's entrance, knowing that although they failed to reach their goal—finding the way out of Sump Four—they lived to tell their story.

Christy Harrison makes her way through a fissure passage in the Beginner's Luck area of Glenwood Caverns.

undiscovered caves, or regions within caves that may have extensions. Yet, it is through determined investigations and luck that cavers are able to enter rooms and chambers never before seen by humankind.

The Dangers of Caving

A great deal of personal risk accompanies cave explorers on their forays underground. Caves can be dangerous places, even to those experienced in underground travel. Hazards include loose and shifting rocks, tight crawlways and fissures, deep pits, fast-moving cold water, and even "bad" air. The complexity of some caves can be a danger, with passages looking alike on multiple levels. The darkness of caves is always troublesome, as the loss of light means a long wait for rescue at the minimum and perhaps death if help does not arrive.

Hourglass Cave, a restricted cave in White River National Forest, is nationally known for the discovery of a skeleton of an early Native American in 1989. The prehistoric explorer quite possibly became disoriented in the complex maze and could not find his way out. His torchlight growing dim, he likely waited for help, but it did not arrive for eight millennia.

In 1908, a group of amateur explorers from Kansas were spared the same fate when exploring Centipede Cave near Manitou Springs. Their candles extinguished in a mad rush to put out a woman's burning hair, the group languished for hours in the jumbled rock of the cave's entrance region. Fortunately, one member happened to see the passing light of a carriage returning to the city that evening, allowing them to crawl back to the surface.

In recent years, numerous accidents have been reported in popular Fulford Cave south of Eagle. Though most are minor, a few have been life-threatening. Fortunately, alpine and specially trained cave rescue teams have been quick to provide assistance to the injured.

The danger of injury or death remains in all caves, even for strong, experienced Colorado cavers like Bruce Unger of Northglenn (see opposite). Since the early 1970s, the National Speleological Society has cataloged cave-related injuries and deaths each year. Sadly, accidents continue to happen underground, predominantly among underequipped and untrained explorers. Always take caution when venturing into caves.

across Colorado arrived by the carload, hopeful of seeing a cave that only a fortunate few were allowed into for the last quarter-century. With surveying and commercial development proceeding throughout the winter months, cavers soon realized that their predecessors in the late 1950s and early 1960s had not examined each and every lead.

Just off the historic visitors' trail in the cave's upper level, a survey team discovered a blowing lead. With a little digging, the hole eventually dropped into the Polar Bar and 1,000 feet of previously unknown cave.

Another tight passage near The Barn required the removal of a single rock to continue. This passage led into a stunning chamber known as Beginner's Luck, containing the cave's largest known lakes. At the Canyon, a seldom-visited upper-level passage, a short dig provided access into Discovery Glenn, a splendid new chamber filled with a variety of colorful dripstone formations.

There's no set method for finding new caves or passageways. Geologists can suggest areas most likely to contain

Beyond the Gypsum Halls of lower Glenwood Caverns, Evan Anderson and his survey team finally call it a day. During the last few hours, they surveyed more than 400 feet of passage, much of it tight and twisty. Beyond are the large rooms and hallways, beckoning each to drop his survey gear and explore.

Fortunately, they ignore temptation. Mindful of their responsibilities to the cave and the survey project, the cavers await Evan's decision. Aware of the effort and time required to produce an accurate map of the cave's seemingly random wanderings, Evan closes his book. He calls his colleagues back, as it's late and dinner calls. There is plenty more for another day.

Goose Creek Tragedy

On a hot August afternoon in 1980, Colorado cavers lost a valued friend to a deceptively challenging cave. Donald Bruce Unger of Northglenn was a tough caver with significant experience in caves around the world. His caving led him throughout the western United States and Mexico to caves more difficult than any in Colorado, as well as to caves in lands as distant as New Zealand and New Guinea. In Colorado, Bruce led the exploration of many caves, including the July 1980 discovery of the White River Plateau's Thursday Morning Cave, a cave filled with difficult chimneys and tight squeezes. At only 30 years of age, Bruce was one of Colorado's strongest cavers with a reputation for hard caving.

Cavers have long been interested in the unique granite caves of the Lost Creek Wilderness, a rugged and remote area southwest of Denver. Completely dark, the caves have ceilings consisting of boulders and even land bridges of soil and trees. In the late 1800s, the Denver Water Board tried to build a dam and reservoir in one of these land bridges, but was stopped by Goose Creek Cave, the longest known granite cave in the state.

On August 8, 1980, cavers incompletely understood the underground wanderings of Goose Creek Cave. Though the stream entrance and exit were well known, the central cave, as well as the concrete remnants from the 19th-century dam-building efforts, were as yet undiscovered. Bruce arranged a weekend expedition with three other experienced cavers to explore upstream along the stream from the lower entrance.

Although water conditions in the cave were not particularly high, Goose Creek was still a potential danger. The four took care in selecting a route through the cave, avoiding the water by chimneying whenever possible. Finally, when the walls became too steep and slick, the group dropped into the stream. Though each wore a full wetsuit, the cold water was still a shock. The water ranged from chest-deep to over the head, making travel upstream slow.

At a chute not far into the cave, Bruce took the lead. He had complained earlier that his boots seemed slick on the wet granite rock. Moving with exceptional care, he straddled the stream above the chute and inched toward an upper pool, where the water flowed more slowly. Unexpectedly, he slipped on the rock and fell into the stream, his foot getting stuck in the chute.

His companions quickly came to his aid. Providing Bruce support in the forceful flow of Goose Creek, they tried to lift and pull him free. Unfortunately, he could not budge, as his boot was wedged in a slender underwater crack. As the minutes passed, the situation became more desperate as each began to tire from the relentless force of the water. Knowing Bruce's life depended on their actions, the other cavers worked with deliberate care in attempting to pull him free.

Tragically, one caver after another washed down the chute, their carbide lamps extinguished, leaving Bruce alone to face the water. By the time the group was able to climb back, he had been completely underwater for several long minutes. Facing a grim reality, the cavers realized that their friend was dead and there was nothing they could do to help.

The body recovery took two days of effort. A solemn group of Colorado cavers came to Goose Creek Cave to release Bruce's lifeless body from the chute that held it so tightly.

Although several individuals previously lost their lives in accidents in Williams Canyon's Huccacove Cave, including a 1916 suicide, Bruce's sad demise in Goose Creek was the first and only to date of a caver on an organized trip in a Colorado cave.

PHOTO BY RICH MERKELY, COURTESY DAN RIDGEWAY.

Cavers examine the Leavenworth section of Narrows Cave. PHOTO BY CAROLYN ENGLUND CRONK.

Perseverance in Narrows Cave

Modern Colorado cavers are not the first to understand the importance of digging in filled cave passageways to find hidden extensions. In the late 19th century, explorers routinely dug and hammered their way through caves in search of precious minerals and additional passage. Ben Snider and Guy Boyd spent weeks in 1929 excavating the connection between Cave of the Winds and Manitou Grand Caverns.

For good reason, today's governmental authorities require detailed dig proposals before the first shovel is turned. These requirements by the Forest Service and the Bureau of Land Management help protect cave resources, as disturbing cave sediment can destroy valuable archaeological and paleontological sites. Digging can also disrupt cave airflow, changing the underground environment and disrupting native cave life.

In private caves like Cave of the Winds and Glenwood Caverns, cooperative caver projects ask that digs be registered and that necessary precautions be taken to protect scientific findings. Often, dig project participants are asked to restore a dig site after completion and install environmental protection gates in the event of a discovery.

Generally, most cave digs in Colorado are short affairs, lasting at most a few months or a year. Cavers will usually spend one day a weekend digging and the other day recovering from their efforts.

An exception to the rule of speedy breakthroughs is Narrows Cave, a mile-long cave in lower Williams Canyon. When Larry Fish and I first began excavating the seven-foot-long cave in 1983, we had no idea it would be 12 years before the first discovery of any length or importance.

Digging followed the cave's entrance passage south into the mountain. Unlike other Williams Canyon caves, the fill was not the usual red sticky clay, but tightly packed sand and gravel. Geologists report that the Williams Canyon stream once flowed into the cave, plugging its entrance passage with debris.

On occasion, digging exposed an air pocket or dome. A small chamber named the Oval Office was found prior to the 1984 presidential election. Though it was only 10 feet long and four feet wide, its discovery gave rise to

The entrance to Narrows Cave perfectly frames Cave of the Winds. PHOTO BY CAROLYN ENGLUND CRONK.

Randy Reck uses a wheelbarrow during the excavation of Zephyr Avenue in Narrows Cave, circa 1995.
PHOTO BY CAROLYN ENGLUND CRONK.

Dan Sullivan on the other side of a tight squeeze in Narrows Cave. PHOTO BY CAROLYN ENGLUND CRONK.

optimism that a major find would be forthcoming. Excavating the floor leading south from the Oval Office, cavers uncovered old flowstone and a small stalagmite buried in the stream channel. Certainly, this cave had promise.

The growing length of the cave gave rise to innovative digging techniques. New methods of transporting debris to the entrance evolved, from passing buckets to dragging snow sleds and baby bassinets on ropes to pulling wheeled carts holding two five-gallon buckets. Cavers even discussed building an electric train that would follow a fixed wooden track to the entrance.

When wheeled carts became difficult to use, cavers turned to full-scale digging. Electric jackhammers and wheelbarrows were brought into the cave, as were electric lights running off a gasoline-powered generator, giving Narrows the appearance of a mine rather than a natural cave.

A hiatus of a few years during the early 1990s convinced cavers to return to the digging style of an earlier era. From the end of the excavated walking passage 150 feet into the cave, wheeled carts and hand excavation allowed cavers to reach an apparent dome. In March 1995, cavers uncovered a few

hundred feet of dry passageway off the top of the dome, the first in Narrows not requiring excavation.

From this Inquisition Dome 250 feet into the cave, dedicated cavers continued to dig south, and an additional 250 feet was opened to yet another ascending passage. In April 1996, this sand- and cobble-filled dome was excavated and opened into an extensive and complex level of passageways.

Finally, Narrows Cave lived up to its promise, with a variety of corridors, chambers, and squeezes awaiting exploration. One passage, called Idiot's Paradise, led up 57 feet to a third level characterized by subway-sized passageways. Another passage was followed through a complex jumble of rock breakdown called Shakedown Street into the Subluna Passage and Spring Canyon. Other passages, such as Great Western and Southern Scoops, suggest that not all is yet found in Narrows.

In all, an estimated 500 tons or more of debris have been excavated from Narrows Cave—far more than any other cave-digging project in Colorado history.

The Legacy

Intently examining the photograph in his hands, Marc Hament shakes his head.

"No," he says, motioning to the ledges at our right. "We're a little too high. We need to drop down a little."

Standing on the steeply sloping wall of lower Williams Canyon north of Manitou Springs, I feel my feet slip a little on the loose dirt. Wherever the photograph was taken, the photographer certainly flirted with the possibility of an unexpected quick trip to the canyon floor.

Marc and I are seeking the exact location of where this photograph of the Manitou Cave entrance building was taken. Cave historian Donald G. Davis came upon this single black-and-white image during his exhaustive research into the history of Cave of the Winds.

Surprisingly, little is known about the history of caves and caving in Colorado. Even archives at Cave of the Winds, the state's most continuously operated commercial cave, are incomplete.

🝙

A good example of this lost legacy is Manitou Cave. Open to the public from 1911 to 1913, the cave's history has almost completely disappeared. Searching Colorado State Historical Society newspaper microfilm records in 1987, I discovered a series of forgotten *Manitou Journal* articles about the cave. Opened as a competitor to Cave of the Winds, Manitou Cave attracted many eager visitors its first summer. Unfortunately, visitors decreased over the next two years, even though the cave was only a 15-minute walk from Manitou's resort hotels. By late 1913, little mention of the cave appears in any literature, suggesting it may have already closed and was awaiting its inevitable sale.

Manitou Cave was probably not much more than a minor irritant to managers of the "Wonderful Cave of the Winds," as advertisements called the famous attraction. Already, the commercial cave had outlasted two other rivals: Colorado's first commercially operated cave, the historic Mammoth Cave near Narrows Cave, and Manitou Grand Caverns on the west side of Temple Mountain.

Mammoth Cave opened in March 1875, after limestone quarrymen with a limekiln near Narrows Cave discovered some of their rock disappearing into a natural crevice. As Colorado's first commercial cave, and one of a handful open in the country, Mammoth Cave initially drew many visitors from Manitou's Cliff House Hotel and other lodges. However, by 1880, Mammoth Cave closed following years of declining visits and growing interest in the recently opened Cave of the Winds. Unlike Cave of the Winds, Mammoth was virtually without decoration and required visitors to follow a strenuous loop tour route involving considerable crawling and traversing of seemingly bottomless pits.

Cave of the Winds, acquiring the famous Manitou Grand Caverns from developer George W. Snider through 1890s

court actions, closed it after the 1906 visitor season. While the Grand Caverns was arguably better decorated, the combined attractions' management decided the cliffside view from the Cave of the Winds' entrance building was superior to that from the Grand Caverns' entrance building. When electric lights were added to Cave of the Winds in July 1907, the management stated the Grand Caverns would be electrically lit and reopened at an unspecified later date. The caverns are still awaiting electric lights today.

While at times ruthless in its business dealings, Cave of the Winds has succeeded like no other Colorado commercial cave. In 120 years of public visits, the cave has remained profitable. Other commercial caves in the state should be so fortunate.

Alexander's Cave, a western Glenwood Canyon cave exhibited since 1893 as Cave of the Clouds, found paying visitors declining within only a few years. It was originally discovered by railroad surveyors in 1885 and opened to visitors in 1887, but the lack of easy access handicapped attendance. Local schoolchildren and adventurers frequented the cave's few hundred feet of passageways following its closure as an attraction in the late 1890s. Their visits throughout the next century sadly stripped the cave of its beauty.

Glenwood Springs developer Walter Devereux opened a new vapor cave in 1893, adding to his Hot Springs Pool and Hotel Colorado visitor attractions. Located across the river from the original Native American vapor cave, the new cave served as a natural sauna. It remains open to this day as the Yampah Spa Vapor Caves.

Three years later, in 1896, attorney Charles W. Darrow opened "The Cave of the Fairies" high above the city on Iron Mountain. Discovered the previous summer, the cave's extensive passageways and curious helictite displays fascinated visitors. Unlike nearby Cave of the Clouds, reached by a foot and burro trail, a fine carriage road allowed direct access to the Fairy Cave entrance. Within the next four years, the Fairy Cave management would add electric lights to their tour

TOP: The remains of a retaining wall associated with the former Manitou Grand Caverns entrance tunnel tell of the cave's long history as a tourist attraction.
MIDDLE: In this circa 1889 photo, visitors to Manitou Grand Caverns enjoy a view of Pikes' Peak from the entrance building balcony. PHOTO BY HERSOM, COURTESY COLORADO SPRINGS PIONEERS MUSEUM.
BOTTOM: In this circa 1893 photo, visitors pause near a large stalagmite in the entrance passage of Cave of the Clouds. PHOTO COURTESY FRONTIER HISTORICAL SOCIETY.

CAVE
OF THE
WINDS.

DISCOVERED 1880 | NO CAVE IN THE WORLD IS BETTER LIGHTED. VISITORS CARRY NO LAMPS OF ANY SORT. | ALTITUDE 7475 FT.

WONDERFUL
DISCOVERED 1880
CAVE OF THE WINDS
ALTITUDE 7475 FEET

TO CAVE OF THE WINDS, MANITOU, COLO   S-254

ABOVE: In this early 20th century view of the Cave of the Winds entrance building, several visitors await their tour.
PHOTO COURTESY CAVE OF THE WINDS COLLECTION.
LEFT: The 1915 opening of Serpentine Drive allowed automobiles to access Cave of the Winds' cliffside entrance building.
PHOTO COURTESY CRONK COLLECTION.

route (being among the first commercial caves in the nation to do so) and excavate a 125-foot-long tunnel to a cliffside balcony called Exclamation Point. Though visitors enjoyed the spectacular canyon view and the cave's curious natural decorations, a substantial drop in attendance closed the attraction by 1912. Fairy would not reopen for public tours for more than 85 years.

Most likely, close proximity to a major summer resort at Manitou Springs helped Cave of the Winds survive the lean years surrounding World War I. Following the 1913 purchase of Manitou Cave, the owners reorganized the attraction as a limited corporation to bring financing for needed improvements. Topping the list of improvements was the 1915 construction of the $8,000 Serpentine Drive highway, leading down along the spine of the ridge from the cave to Manitou

Springs. The July opening of this switchbacking road allowed the city to authorize the travel of automobiles up Williams Canyon and down Serpentine Drive.

Four months later, automobile entrepreneur Henry Ford and inventor Thomas Edison visited the cave while touring the country. Ford was delighted that his new automobile managed the steep Williams Canyon road to the Cave of the Winds entrance, while Edison showed considerable interest in the tour's electric lighting.

Mining: The Key to Cave Discoveries

Historical research can provide many insights into the early days of Colorado caves. Not surprisingly, several of Colorado's best-known caves were found during the 1880s and 1890s,

Whispers of an Ancient Time

In the 15 to 20 centuries that humankind has called the Colorado Rockies home, caves have served a variety of purposes.

In the flat tablelands of Colorado's southwestern deserts, large sandstone shelter caves allowed the Pueblo Indians to build elaborate dwellings protected from the weather. These stone and mud structures, occupied for about 200 years until 1300, presumably also shielded their inhabitants from raids by nomadic groups. Modern archaeologists continue to study these ancient cities, now protected in Mesa Verde National Park and the Ute Mountain Tribal Park. It is believed the cities were abandoned over several decades following years of drought that withered crops and dried up springs and streams.

A solutional dolomite cave (an underground passage developed in dolomite rather than limestone) in Dinosaur National Monument, which straddles the present Colorado–Utah border, may have been a defensive hideout for early Native Americans. In 1981, Colorado cavers exploring Cave of the Logs discovered 18 logs up to 10 feet in length randomly arranged about 200 feet inside. One of the discoverers, Donald G. Davis, wonders if a study of the rat midden that buries the logs might establish an age ranging in the thousands of years. Curiously, another small Colorado cave, BAD Cave near Dotsero, has a similar arrangement of buried logs.

Two other limestone caves near Dotsero, Sweetwater Indian Cave and Ute Indian Pictograph Cave, boast extensive pictograph panels from prehistoric Native American artists. Using natural rock pigments from area iron deposits, these artists depicted men, animals, and abstract art in their galleries. Sadly, the better-known Sweetwater Indian Cave has fallen victim to extensive vandalism in the last few decades as visitors have added their own painted names and even used the art as shooting targets.

Native Americans frequented the original Vapor Cave at Glenwood Springs. Located along the south bank of the Colorado River at the west end of Glenwood Canyon, this cave was blocked by the construction of a railroad in 1887. The cave's natural steam and hot springs baths attracted members of many area tribes, who considered it a sacred site.

Likewise, the Manitou mineral spring on Colorado's eastern slope was a popular destination of Native Americans. Marked with a brightly colored travertine mound on the north bank of Fountain Creek, the spring and its effervescing pools were sacred.

According to stories and legends passed down by spoken word, tribes knew of Cave of the Winds in the canyon north of Manitou's mineral spring but did not venture inside. Certainly, the dirt fill that blocked the cave within a few feet of its entrance kept anyone from exploring. However, the Apache Indians believed the winds that blew through the large natural arch at the cave's entrance signified that it was the home of their Great Spirit of the Wind. They held that this spirit caused whirlwinds on the plains to the east and the frequent winter downslope windstorms we call chinooks. Tribal authorities warned that anyone who dared enter the cave from which the wind blew would become twisted in mind and spirit.

Hailing from the mountains west of Manitou, Ute Indians who visited the spring at Manitou also knew of Cave of the Winds. In 1914, Ohio poet James Henry Williams published a book on the Ute legends of the cave he had learned. The Utes told of a great Chief White Wing and his daughter, Juanita, who took refuge in the cave's spacious entrance in a futile effort to hide from the approaching Europeans. Awaiting rescue from the mighty warrior Wampuno, Juanita's lover, the two died from hunger and thirst. Following a vision of his lover, Wampuno hurried to the Cave of the Winds entrance, where he found their lifeless bodies. The warrior's grief-stricken cries became the winds that blow through the arch in memory of Chief White Wing and his daughter.

ABOVE: Large sandstone shelter caves house ancient dwellings at Mesa Verde National Park. PHOTO BY BILL BONEBRAKE. RIGHT: According to Ute legend, warrior Wampuno mourns his lost lover at Cave of the Winds.

when prospectors wandered the mountains in search of gold and silver.

One of the first reported discoveries of a cave in Colorado's high country came in August 1879, when prospectors near Ouray found the 350-foot-long Canyon Creek Cave. Area mines later reported intersecting natural caves, including the National Bell Mine near Red Mountain Pass and the rich American Nettie Mine. According to published reports, the American Nettie intersected miles of natural passageways.

Prospector J. H. Yeoman, searching Marble Mountain in the Sangre de Cristo Mountains, came upon the entrance of White Marble Halls in September 1880. Undoubtedly, Yeoman also visited Spanish Cave higher on the mountain, for Moon-milk Cave across from its entrance contains Yeoman's distinctive signature on the wall.

To the east of Marble Mountain, miners near Beulah found Whistling Cave in the early 1880s. Although the cave was reported to contain more than 800 feet of passage, area residents sealed its entrance by 1900 as it was considered dangerous. A cowboy found another nearby cave, Hangman's Cave, during an 1882 cattle drive. Apparently, a cow fell into the cave's entrance, requiring rescue by the cowboy and a Beulah prospector.

North of Cañon City, a limestone quarry unexpectedly revealed the jagged pit entrance to Marble Cave in 1880. While there is no evidence that the cave was ever operated as an attraction, by 1881 it was a fashionable destination of area residents who explored its passageways armed with candles and lanterns.

The mining camp of Carbonate on the central White River Plateau was Garfield County's first county seat. Reached by rough stage roads from the east and south, the isolated camp saw many cold and snowy winters. However, during the summers, warm, sunny days and flower-filled alpine meadows inspired residents to hunt for minerals and enjoy the outdoors. Along the rim of scenic Grizzly Creek, picnickers apparently frequented Bair Cave, a comparatively undemanding cave with two entrances. Numerous historic signatures and dates from the early 1880s grace the cave's western walls and ceiling.

Prospectors investigated other White River Plateau caves during this era. The 1888 signature of F. A. Kimball adorns Kimball Cave in Tie Gulch near Glenwood Canyon. Premonition Cave in Deep Creek contains historic signatures, as does French Creek's Skeleton Cave.

A ruined cabin marks the old mining town of Fulford.

A prospector named Hooper apparently discovered in December 1891 one of the plateau's more popular caves. Telling his story to the citizens of nearby Meeker, he found that they showed little interest in his fantastic tales of a cave with an underground river. By the late 1920s, however, increasing curiosity about Spring Cave encouraged the White River National Forest to install a sign along the White River identifying the trail to the cave. In September 1931, ranger Earl Ericson produced from field notes the cave's first known map.

Near the bustling mining town of Fulford in Colorado's central mountains, E. Maxfield, R. C. Koontz, and other partners seeking valuable minerals instead discovered a cave in late 1892 on their Shamrock Claim. Those who ventured into its many winding passageways considered the Shamrock Cave, now known as Fulford Cave, a rival of Kentucky's great Mammoth Cave. By May 1893, Maxfield found that the visitors who wished to tour his cavern often interrupted his mining activities. Hoping to find his fortune in cave tourism, he installed ladders and made other improvements, and the visitor numbers increased. It is claimed that Maxfield even considered building a hotel to take advantage of the interest in the cave.

Later that summer, William Henry Hubbard, his brother Charles, and brother-in-law Griffith Jones came upon the multiple entrances of Hubbard's Cave on Glenwood Canyon's south rim. Discovered by the men and their dogs while hunting for game and minerals, the cave contained signs of previous use as a Ute Indian campsite. Complete with fire pits, arrowheads, and primitive utensils, the camp appeared to have been occupied for many years before abandonment. The three men reported their discovery to the citizens of Glenwood Springs, and a rough wagon road was built to a point near the cave, allowing access for summer visitors.

In many of Colorado's mining districts, the discovery of caves often brought good fortune. Ore deposits were sometimes located in and around natural vugs and fissures. In the mines of Aspen and Leadville, 19th-century miners discovered that the Leadville Limestone was mineralized with rich deposits of gold and silver.

On occasion, miners excavating tunnels discovered caves with no natural entrance. To the south of Hartsel, a group of miners following an ore vein in the limestone bedrock broke into Porcupine Cave in about 1884. While the privately owned cave was briefly considered for commercialization and even national monument status in 1925, the lack of natural decorations or even spacious chambers probably discouraged any active pursuit of paying visitors. Today, this dry and dusty cave is the site of continuing paleontological exploration by several museums, including the Denver Museum of Nature & Science. Since the mid-1980s, scientists have found one of North America's richest sites of Pleistocene mammal bones here.

Discovered by miners around 1884, Porcupine Cave bears one of North America's richest deposits of Pleistocene mammal bones.

In 1893, miners seeking lead and silver broke into a descending cave along an alpine creek near Fairplay. They recorded their discovery at the county assessor's office, naming it Radford Cave. The abandoned mine entrance's timbering collapsed around the turn of the century. Miner Jack Moran reopened it in 1938, but his timbering collapsed in 1975, closing Radford Cave (now known as Cave Creek Cavern) until the moment in September 1996 when cavers successfully reopened it.

Near the town of Fulford, miners A. D. McKinsey and Joe Good intersected a large natural cavern in 1895. Although this cave had contained no valuable minerals, a large chamber provided a handy dumping place for tailings from the ongoing Bower Tunnel project, which was excavated until 1902. The tailings provide a relatively easy route down the vertical shaft leading to the bottom of "Devil's Den," as cavers named it upon rediscovery in August 1964.

Organized Caving Brings More Findings

Not surprisingly, the first efforts to locate, identify, and explore all Colorado caves began with the coming of the National Speleological Society (NSS) and organized cavers in 1951. Except for the late 19th-century activities by the owners of Cave of the Winds and Manitou Grand Caverns to locate new caves for possible commercial development, the chartering of the Colorado Grotto (the NSS's Denver chapter) in November brought together the few cavers in the state for the first time.

At first, the cavers had their hands full. As the Colorado Grotto was the only organized group between the Mississippi River and the West Coast, its members were invited to explore and document caves ranging from Montana to New Mexico. In Colorado, the citizens of Glenwood Springs asked the cavers to fully explore Hubbard's Cave. The town of Cody, Wyoming, requested that the Colorado Grotto visit Frost

OPPOSITE: June Harris peers into the east entrance of Hubbard's Cave; discovered by a hunting party in 1893, the cave's main entrance is likely a former Ute Indian campsite.

Anxious to have cavers assess Hubbard's Cave for possible commercial development, a group of Glenwood Springs citizens accompanied Colorado Grotto members on this 1952 trip. PHOTO BY GLENN POLLARD, COURTESY JOHN STREICH COLORADO GROTTO COLLECTION.

On this 1932 Colorado Mountain Club expedition to Spanish Cave, a caver belays an unseen climber below, while another caver shines his flashlight. PHOTO COURTESY DENVER PUBLIC LIBRARY, WESTERN HISTORY COLLECTION.

Colorado Grotto cavers follow the abandoned historic Cave of the Fairies tour route in present-day Glenwood Caverns, circa 1955. PHOTO BY JOHN STREICH, COURTESY JOHN STREICH COLORADO GROTTO COLLECTION.

Cave, a former national monument. The National Park Service invited Colorado cavers to visit New Mexico's Carlsbad Caverns National Park to explore and document some of the park's smaller caves. Among the caves explored was a little cave they called Lechuguilla. On a dig project some 35 years after the club's initial exploration, Colorado cavers would discover the hidden inner cavern, revealing Lechuguilla as one of the world's great caves.

Curiously, until the 1980s, Colorado cavers were not allowed to explore off the commercial trail in Cave of the Winds. Manitou Grand Caverns, connected by a passage excavated in 1929, was off-limits. Cavers were permitted, however, to explore and survey the former Mammoth Cave, now known as Huccacove, and the former Manitou Cave, once named Pedro's Cave.

Another favorite in the 1950s and 1960s was Spanish Cave, where caver Donald G. Davis and the Southern Colorado Grotto explored well beyond the yawning entrance shaft that captivated 1920s and 1930s Colorado Mountain Club and Forest Service explorers. Donald and others discovered that Spanish Cave isn't as mysterious as legend claims. No evidence was ever found of previous visits by either Native Americans or Spanish conquistadors.

The Colorado Grotto also explored Fairy Cave and Fulford Cave in the 1950s, spending time pushing small and overlooked passageways. By the early 1960s, noteworthy new rooms and passages had been discovered, including the Canyon, The Barn, and King's Row in Fairy, and the Cathedral Room and Attic in Fulford.

Spring Cave, near Meeker, also garnered attention during this era. In 1958, cavers discovered an unknown upper level

Spanish Lessons

For nearly a century, Colorado newspapers have told of a fabulous cave in the Sangre de Cristo Mountains. Known as La Caverna del Oro, "the cave of gold," this cave was allegedly explored more than four centuries ago by Spanish conquistadors. Exploiting the Native Americans who called these rugged lands home, for God and Spain the cruel soldiers forced them to dig gold by flickering candlelight. Death was the only release from this slavery in the cave's gloomy depths. As legend reports, rebellion eventually drove the Spanish back into Old Mexico without their gold. Those who remembered the cave's terrible past warned future visitors away.

A 1950s Colorado Grotto expedition member enjoys a meal at the bottom of Spanish Cave. PHOTO BY JOHN STREICH, COURTESY JOHN STREICH COLORADO GROTTO COLLECTION.

A participant in the 1932 Colorado Mountain Club Spanish Cave expedition. PHOTO COURTESY DENVER PUBLIC LIBRARY, WESTERN HISTORY COLLECTION.

Despite its intriguing name, Spanish Cave has absolutely nothing to do with Spanish explorers. Since its discovery by prospector J. H. Yeoman in 1880, Spanish Cave has been more closely associated with the hardy German settlers of the Wet Mountain Valley than it has with ancient Spanish conquistadors, gold diggings, slaves, and hidden treasure. Yet, fueled by the painted Maltese cross at its entrance, whispered rumors, and sensational newspaper articles, the cave's legend continues to this day.

Denver caver Donald G. Davis is perhaps Colorado's foremost cave historian. His research has brought to light much of the long history of Cave of the Winds and Manitou Grand Caverns. A caver for more than 40 years, Donald has been at the forefront of Spanish Cave exploration.

Before the 1960s, little was known of Spanish Cave beyond its deep interior shaft called The Jug. Expeditions to the cave in the late 1920s and early 1930s by Carl Blaurock and the Colorado Mountain Club proved the abyss held no gold, chained skeletons, or ancient tools as had been claimed in lurid newspaper accounts. Rather, Spanish is a cold, wet cave—a danger to those who believe stories of untold wealth hidden within.

In the late 1950s and early 1960s, Colorado Grotto began exploration of the cave, one of the highest known in the nation. Among the explorers was Donald, who discovered an overlooked passage leading to the main part of the cavern. During his many expeditions, Donald helped explore two-thirds of the cave known today. Donald believes there is no evidence that man previously entered, much less mined, the cave.

Donald and the cavers who explored, surveyed, and studied Spanish Cave concluded that the Pennsylvanian-age limestone in which the cave is found lacks any precious metals. The cave most likely formed not from the ascending hot mineral waters that sometimes deposit rich lodes of gold and silver, but as a result of cold, descending glacial meltwaters, which dissolve calcite from certain types of rock, including limestone, marble, and dolomite, to carve out a cave. With several thousand feet of surveyed passage and an upper entrance at nearly 12,000 feet in elevation, Spanish Cave is unlikely to have extensive unknown sections awaiting discovery. At nearby White Marble Halls, also explored since 1880, cavers have been frustrated in finding passageways leading closer to Spanish. Even in lower Spanish, at the bottom of The Jug beyond the so-called Gold Diggings, efforts to extend the known cave by following a small underground stream have met with disappointment.

Although the famous Maltese cross at Spanish Cave's entrance has faded considerably over the decades, careful studies indicate it is actually a German cross. One of the German immigrants who lived in the valley below probably painted the cross sometime in the late 19th century. A smaller faded cross of a similar design also marks the entrance to White Marble Halls, along with elegant English-language graffiti stating "400 feet to the White Marble Halls."

Climbing the long steel cable ladder in Spanish Cave's The Jug. PHOTO BY JOHN STREICH, COURTESY JOHN STREICH COLORADO GROTTO COLLECTION.

Folklore notwithstanding, Spanish is an interesting and complex cave. The challenging journey from the cave's upper entrance to its lower entrance is a favorite of cavers who are experienced in technical ropework. Spanish is not a cave to be taken lightly, however, as even highly skilled cavers have suffered serious injuries from falls, loose rock, and other hazards. The frigid temperature and steady chimney-effect wind blowing from one entrance to another can afflict any incapacitated caver with hypothermia. Fortunately, cavers have so far managed self-rescues from the cave and have brought the injured successfully to hospitals for treatment.

While the legends of Spanish Cave have proven false, cavers report one mysterious occurrence during their many years of exploration. On certain evenings, while exiting Spanish Cave late at night, sharp-eyed cavers have spotted a curious torchlight high on the slopes of Marble Mountain. Cavers have followed the flickering light for several minutes as it moves quickly across the darkened alpine landscape. Eventually it disappears, leaving them wondering if they simply imagined the sighting or if someone—or something—walks alone on the high mountain.

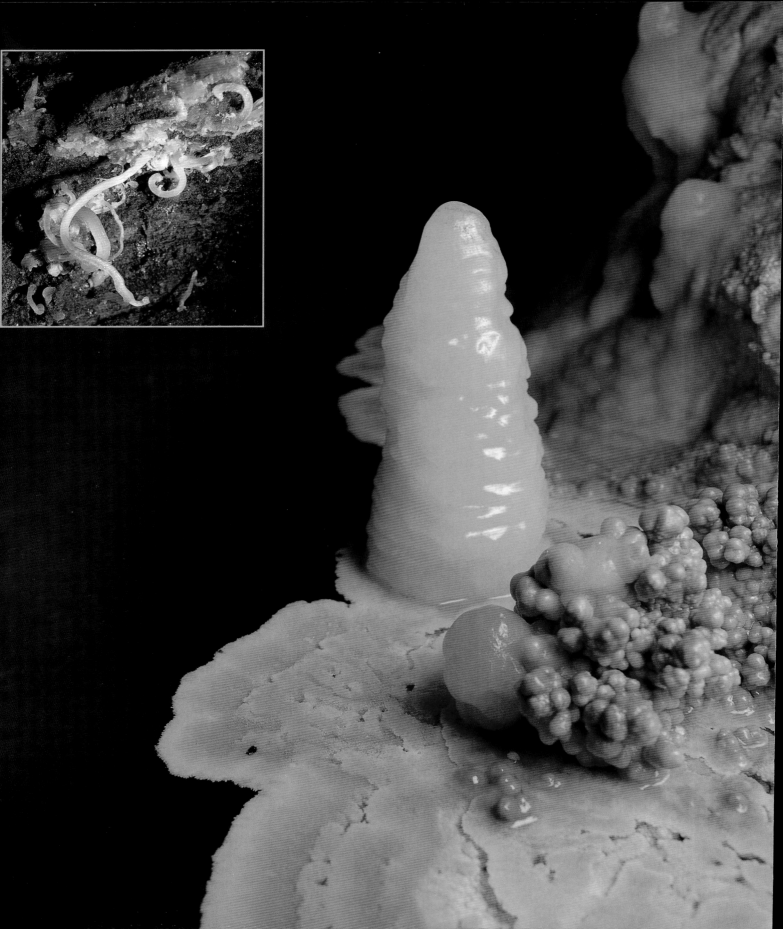

leading to thousands of feet of passage. A Thanksgiving 1962 expedition by the new Colorado School of Mines Student Grotto fully explored and surveyed this new section, which includes the state's most impressive stream gallery.

Improved Colorado highways helped bring the White River Plateau closer to Denver metropolitan cavers during the late 1960s and early 1970s. The opening of Interstate 70 west of the city provided easier and timelier access to the plateau's vast array of limestone caves. In August 1968, Colorado School of Mines cavers John Pollack and Paul Westbrook came upon the entrance to Groaning Cave while searching for buffalo bones. Though the cave was well known for many years by area ranchers and hunters, no one had carefully examined the back of its entrance hall. To the cavers' great surprise, a short dig in soft dirt immediately led them into an extensive series

Alan Williams examines pristine draperies on the ceiling of Groaning Cave.

Dan Sadler carefully straddles a pool of water in Groaning Cave's CSU Passage.

of parallel passages. By the end of 1969, surveying in the cave extended Groaning beyond the length of Fulford, Spring, Hubbard's, and Fairy, the state's longest surveyed caves.

Groaning's discovery helped revitalize cave exploration in Colorado for the last quarter of the 20th century. Proving that not all had been discovered, Groaning's miles of new passage encouraged intensive exploration of the White River Plateau and other prime limestone areas. While exploration and survey continued in Groaning, bringing the cave to 10 miles of passage in 2000, other cavers explored the plateau's limestone cliffs and exposed bedrock. Many significant caves were discovered as a result, including Fixin'-to-Die, Premonition, Thursday Morning, Twenty Pound Tick, Summer's End, Spectre, and Buffalo. Significant discoveries in the Lime Creek and Fulford areas south of Eagle, along with new finds in the San Juan Mountains and Taylor Park, helped establish the 1980s and 1990s as the most prolific decades in the history of Colorado cave exploration.

Cavers returned in 1982 to Williams Canyon for the first time in 20 years, as new management at Cave of the Winds recognized the benefits of organized exploration and study. Not only were new, detailed maps of the caves prepared, but exciting and extensive discoveries were made nearly every

OPPOSITE: An iron oxide-tinted stalagmite, rimstone dams, and botryoids line a pool's edge in Groaning Cave.
INSET: The curving strands of a gypsum flower, rare in Colorado, extrude themselves from a passage wall in Groaning Cave.

Each holding a drop of water, hollow calcite stalactites called soda straws line the ceiling of the CSU Passage in Groaning Cave.

year. New caves like Dilation, Three Hole, Narrows, and Breezeway helped establish the canyon as one of Colorado's most noteworthy. Digging in well-known caves such as Huccacove and Manitou (Pedro's) revealed interesting new sections. Studying the geology of the canyon, cavers even began to imagine linking the caves together someday.

Standing on the canyon hillside overlooking Manitou (Pedro's) Cave, I can see where the Manitou Cave entrance building once stood. While time and at least three great floods have changed the canyon floor considerably in the 84 years since the photograph was taken, the cliffs and contour of the landscape remain largely the same.

Although many mysteries about the former commercial cave have yet to be resolved—even the length of the tour route—I realize that the cavers of this era know much more about this canyon's caves than our predecessors ever imagined. Through continued exploration and scientific study, along with researching historic reports, photographs, and documents, we can use this knowledge to discover rooms, passageways, and even caves that as yet remain hidden.

Much like the miners and entrepreneurs who passed through this landscape before us, we are still learning the potential of Colorado's caves—a never-ending process of discovery.

OPPOSITE: Shelfstone rims lace the edge of a pool in Groaning Cave's Blue Pool Room.

Manitou's Lost Tour

A few hundred yards north of Manitou Springs, cavers have spent many years digging and searching for part of a cave lost since at least 1921.

Thanks to flash flooding in the canyon, Manitou Cave (a.k.a. New Cave and Pedro's Cave) is missing a section once shown on a commercial tour route. Today's wild tours to the cave visit a portion of this old route, complete with an antique hand railing, before venturing off into an extension discovered in February 1994.

Where the lost section is hiding is anybody's guess. Some cavers believe it is southeast of the known cave, on the east side of Williams Canyon. Others believe it is to the south, in a region strongly suspected by geologists to have unknown cave passages.

History records that in July 1911, entrepreneurs J. F. Sandford and R. D. Weir of Manitou Springs opened "New Cave" to public exhibition. Discovered in March 1910 through determined digging and exploration by a force of men, this cave had a more favorable location—closer to the town and its many resort hotels—than did rival Cave of the Winds, located farther up the canyon. In fact, the owners convinced the city council to allow automobiles to travel only up to the New Cave entrance, forbidding travel the remaining distance to the established Cave of the Winds.

Skilled promoters, the owners of New Cave encouraged newspaper coverage of their new attraction. In addition, Weir and Sandford cleverly held a contest among cave visitors for a permanent name—the winner to receive a lot in a new subdivision they were building in Manitou Springs. By September 1911, a new name was selected, Manitou Cave, and the Oklahoma winner received the title to her little lot in the Ouray addition.

Published reports of the era tell of many rooms and corridors in this commercial cave, including passageways unknown today. The stories report sizable chambers, including Convention Hall and a stream gallery with a lake. Nothing of that nature is currently known in Manitou Cave.

Curiously, a single photograph is all that remains of this commercial business despite considerable research to locate old postcards, maps, and

At the Whirlpool Dome in Manitou Cave, pipe railings mark the former tour path.

memorabilia. Quite possibly, the scarcity of historic materials stems from the short term of the commercial operation.

Following a successful first year of operation, Weir, Sandford, and partner D. H. Rupp found themselves financially overextended. In addition to running the commercial Manitou Cave and building a new housing subdivision, they managed two other area attractions: Rainbow Falls and the Red Mountain Incline Railway. In need of operating funds, they leased Manitou Cave and Rainbow Falls in May 1912 to a local businessman who previously ran burros in North Cheyenne Canyon. Although the cave remained open through the summer of 1913, it apparently was never profitable. In October 1913, the owners sold the operation and the surrounding Williams Canyon property to Cave of the Winds.

Having already closed down rival Manitou Grand Caverns in 1907, Cave of the Winds most likely promptly shut down the Manitou Cave operation. The entrance building, located on a platform over the usually placid Williams Canyon creek, was also closed. During the next seven years, the building was leased as a small summer home for a Santa Clara Indian from New Mexico, Pedro Cajeta, who danced for tourists and sold souvenirs and photographs under the name Chief Manitou.

Unfortunately, the housing lease came to an end in June 1921, when a flash flood washed away the former entrance building and caused thousands of dollars in damage to city works and businesses in the path of the floodwaters. Because the entrance to Manitou Cave is located adjacent to the Williams Canyon floor, this flood undoubtedly also infiltrated the cave, filling the passages with deep, muddy water and depositing sand and gravel from the eroded road base inside.

Individuals who visited "Pedro's Cave" in the following years reported that many passages seemed to be missing from their previous visits, granted by Chief Manitou. Additional floods in 1947 and 1965 most likely impacted the cave, completely sealing the historically known extension.

In May 1988, cavers dug open a forgotten connection between the former commercial cave and adjacent Centipede Cave. This crawlway provided access to several hundred feet of passage, including the first known route under the Williams Canyon floor. It also gave geologists additional proof of possible passageways to the south, one of which may be the long-lost tour route. Centipede showed signs of considerable traffic through the years, including old signatures and "out" arrows penciled on the walls, leading to a now-sealed entrance.

In August 1911, a Rocky Ford minister visiting Manitou Springs on holiday took the New Cave tour. He came away impressed enough to write a letter to the owners, who forwarded it to the local newspaper for publication. In it, the Reverend O. L. Orton eloquently speaks of the new operation and the potential for the cave, noting the cave owners "will have the mystery cave of the world."

Until more can be learned about Manitou Cave or a lucky caver stumbles upon the missing tour route, Orton's statement is fittingly prophetic.

The old Williams Canyon entrance, circa 1881-1895. PHOTO COURTESY OF PENROSE LIBRARY.

Subterranean Science

Dark, threatening clouds crowd the once-blue summer sky. To the west, lightning flashes angrily among the high peaks. A moment later, the crackling thunder rolls across the gentle volcanic basalt flows of the San Luis Valley. Splashing through puddles of thick red water that flood the rutted lane through the sagebrush, I drive a little faster, hoping to beat the storm to our destination.

With me are geologists Paul Burger and Mark Maslyn. According to the directions provided by fellow caver Dr. Fred Luiszer, we should soon see the University of Colorado camp along the Colorado–New Mexico border. Now and then, I catch glimpses in my rearview mirror of Paul's grinning face as he braces himself among the tumbling piles of gear in the shell of my pickup truck.

We had traveled all morning from Denver to visit a lava tube less than a mile south of the state border. Discovered in the early 1950s and excavated by a local gentleman who had visions of buried Spanish gold, it is the closest known lava tube to Colorado. Because the basalt flows continue north across the border, the likelihood exists that similar tubes will someday be found within Colorado.

Lava tubes and sea caves are about the only types of caves yet to be discovered within the state. Solutional caves in lime-stone, gypsum, and dolomite are known, as are unusual caves in compacted, dry mud and rock talus. Fracture caves in granite and sandstone have been found, as have tufa caves near Rifle Falls and a snowfield ice cave at Saint Mary's Glacier.

Limestone solutional caves are widespread throughout Colorado. Yet, unlike caves in Missouri or Virginia, Colorado's caves lack the usual solution processes. Geologists have found that most of the state's caves developed through unique circumstances uncommon elsewhere.

During the 1970s, Mark and others studied many western caves, including those in Wyoming and New Mexico as well as Colorado. In most caves, geological and mineralogical clues remain as to their origin. To the surprise of the geologists, a number of western caves do not appear to have developed through a traditional groundwater method. Large deposits of gypsum, sulfur, and other minerals instead suggest that these caves owe their origin to aggressive hydrogen sulfide waters ascending from deep oil fields.

Commonly, in most caves worldwide, groundwater with a high carbon dioxide solution slowly dissolves limestone along rock fractures. This acidic water acquires its carbon dioxide from surface plant life and soil as rainwater drains into the ground.

Hydrogen sulfide caves, however, are the result of highly aggressive water, which literally dissolves away limestone. Sometimes these rising waters carry minerals in solution. In the Aspen and Leadville mining districts, miners discovered that filled cave passageways in the Mississippian-age Leadville

ABOVE: Braided Cave in New Mexico's El Malpais National Monument is a classic lava tube, a type of cave not yet discovered in Colorado.
OPPOSITE: The warm light of sunset bathes Fulford Cave's upper entrance.

Cave Formation in Colorado

Most limestone caves around the world owe their development to descending rainfall, but most Colorado caves are likely the result of rising groundwater from deep oil fields or mineral-rich underground springs. Groundwater rich in hydrogen sulfide (if the water originates from oil deposits) or carbon dioxide (if the water originates from mineral springs) follows fissures in limestone beds, eventually emerging at the surface as springs. Just below the springs, the underground water combines with surface water (rainwater), forming a solution rich in sulfuric acid or carbonic acid. These acids dissolve the limestone, creating large, open cavities. When the groundwater eventually drains away, it leaves caves behind.

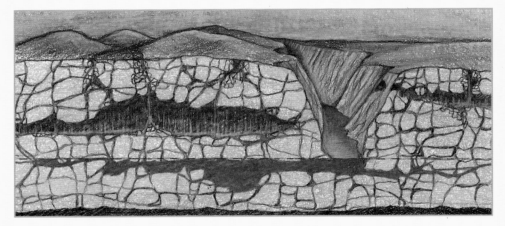

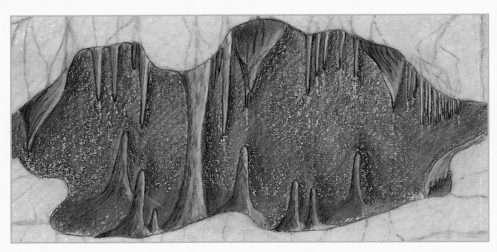

The various formations within caves are derived from surface water filtering down through the limestone beds, dissolving minerals along the way. Over time, as water drips into the caves, the minerals are deposited, creating beautiful cave formations like stalactites, stalagmites, and draperies.

ILLUSTRATIONS BY KIRSTEN MUSTONEN

Limestone often hold fantastically rich deposits of gold, silver, copper, and other metals. Mark's studies show that remnants of ancient Pennsylvanian-age caves, sinkholes, pinnacles, and other features called "paleokarst" were preserved due to mineralization by ascending waters. Near Aspen, Mark found the remains of a huge paleosinkhole a half-mile across. Colorado clearly had many major cave systems in the Pennsylvanian age.

During the 1970s, Donald G. Davis and New Mexico geologist David Jagnow studied the great caves of the Guadalupe Mountains near Carlsbad Caverns National Park. Curiously, many of these large limestone caverns contain thick beds of gypsum, apparently deposited during these caves' development.

Near Lovell, Wyoming, geologists examined two small hot springs caves. These caves also have massive gypsum beds, along with hydrogen sulfide gas escaping from the emerging spring waters.

In the 1980s, geologists decided there is a common link among the Wyoming caves, the caves in New Mexico, and Mark's central Colorado paleokarst. Ascending water, rich in hydrogen sulfide, flowed to the surface and intermixed with shallow groundwater to develop these caves.

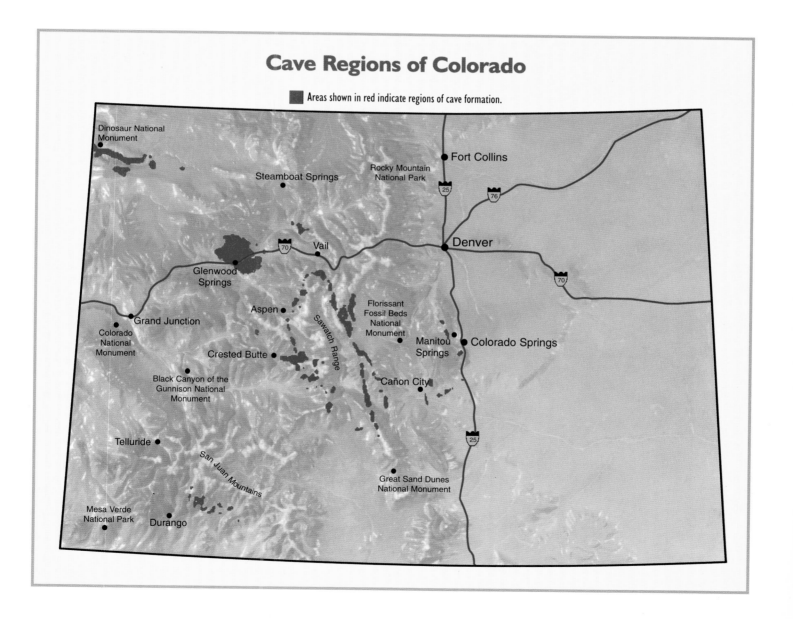

Cave Regions of Colorado

Areas shown in red indicate regions of cave formation.

Dinosaur National Monument

Steamboat Springs

Rocky Mountain National Park

Fort Collins

25 76

Vail

Glenwood Springs

70

Denver

70

Aspen

Sawatch Range

Florissant Fossil Beds National Monument

Manitou Springs

Colorado Springs

Grand Junction

Colorado National Monument

Crested Butte

Cañon City

Black Canyon of the Gunnison National Monument

Telluride

San Juan Mountains

25

Great Sand Dunes National Monument

Mesa Verde National Park

Durango

A tour group in Manitou Grand Caverns' Grand Concert Hall investigates the cave's geology, which has been studied since the late 1800s.

With this important discovery, cavers and geologists began applying the hydrogen sulfide model of development to many western caves. Yet, for other numerous Colorado caves, the model fit no better than the traditional descending groundwater model.

Cave Formation Processes

Mixing Waters

Considered by others to be a hydrogen sulfide cave, Cave of the Winds and its origin intrigued Fred Luiszer. As a University of Colorado geology graduate student, Fred believed that evidence suggested another geologic history. The Manitou mineral springs, effervescing with carbon dioxide, served as clues. At Cave of the Winds, and in distinctive surface outcrops in Williams Canyon, colorful clays and minerals suggest the springs played a key role in the development of the caves. Indeed, it is widely held that the springs are dissolving new caverns below the city today.

Historical investigations of the cave's geology date back to 1893, when Colorado College Professor William Strieby studied the cave and the springs. Other accomplished geologists also examined the cave and canyon, including George Finlay in 1906, J. Harlan Bretz in 1942, and George Morgan in 1950. Until Fred's investigation in the late 1980s and early 1990s, Strieby's ideas were ignored and forgotten. Surprisingly, his conclusions were nearly identical to Fred's, though it took modern science to provide proof.

In his studies, Fred initially examined the clays found within Cave of the Winds. After building a mechanical coring apparatus, he drilled deep into the clay floors of several chambers, including the Grand Concert Hall. Up to 30 feet or more deep, these clay floors hold a fascinating record of the wanderings of our planet's magnetic poles. Through a process called magnetostratigraphy, Fred was able to read the fossil magnetic orientation preserved in the iron-rich sediments. And through numerous reversals of the north and south magnetic poles, Fred determined that the sediment dates back 4.3 million years, about when the cave was first developing.

Analyzing the colorful sediments with powerful university equipment, Fred identified arsenic, lead, and other minerals and concluded that the ascending water from the Ute Pass Fault was their source. Because the springs at Manitou are only slightly thermal (that is, they are not considered hot springs), Fred decided that the ascending spring water, rich in

OPPOSITE: Williams Canyon's colorful clays and minerals suggest that ancient spring waters dissolved Ordovician-age Manitou Limestone to form the caves near Manitou Springs.

minerals and dissolved carbon dioxide, mixed with shallow groundwater to form a highly aggressive carbonic acid. This acidic water easily dissolved the limestone along fractures before emerging at the surface as springs. The curious surface features of dolomite and other colorful sediments found throughout Williams Canyon are likely former spring basins of the current Manitou springs.

Later, as Fountain Creek and other surface streams eroded deeper into the mountain, forming the canyons we know today, the caves drained of the acidic water and allowed the slow mineral deposition of dripstone and flowstone to begin. Caves lower in the canyon, such as Manitou and Narrows, have had less time for such decorative features to form. Drier climates also affected the deposition of stalactites, stalagmites, and other cave features.

Colorado's western slope is home to a similarly formed cave. When current Glenwood Caverns owner Steve Beckley began preparations to open the former Fairy Cave as a commercial attraction in 1998, Fred took advantage of the opportunity to learn more about the cave's geology. Once again, though circumstantial evidence suggests that this cave and others nearby might have formed strictly by hydrogen sulfide solution, Fred thinks otherwise. While thermal waters of Glenwood Hot Springs may have played a role in its development, Luiszer suggests that maybe

Hazel Barton in the profusely decorated Lower King's Row, an area visible from the Glenwood Caverns tour trail.

Evan Anderson conducts a survey near the back of Hubbard's Cave, a cave formed by hydrogen sulfide solution.

OPPOSITE: Paul Burger admires a group of beaded helictites near Stone River in Breezeway Cave, probably home to the highest number of these rare formations in the world.

In high spring run-off years, this glacial outwash-carved passage in Spring Cave can fill with water.

the solution process responsible for creating this cave is the same as that at Cave of the Winds. As at Manitou, ascending carbon dioxide-rich waters mixed with shallow groundwater to form an aggressive carbonic acid. This acidic water followed fractures and faults to the surface, dissolving the limestone and forming caves in the process.

HYDROGEN SULFIDE

Massive gypsum beds in Fairy's Gypsum Halls and Glenwood Canyon's Hubbard's Cave provide physical proof that hydrogen sulfide solution played a role in their geologic history. Fred believes that Fairy, Hubbard's, and even Groaning caves all were ancient outlets of the Glenwood Hot Springs and Dotsero Warm Springs. In turn, each lost its spring as the Colorado River continued to erode deeper into the White River Plateau through the millennia.

GLACIAL OUTWASH

Some Colorado caves, including Spring and Fulford caves, are probably the result of cold, aggressive glacial meltwater. Studies of these caves and their surrounding geological environments by Fred, Donald, and others indicate the nearby presence of icy glaciers some time ago. Chemically, cold glacial water can hold far greater amounts of carbon dioxide than can groundwater. Draining into the ground from large valley or cap glaciers, this water aggressively dissolved cracks and crevices in the limestone, forming caves.

Who Lives Underground?

Standing in the darkened chamber beneath the Cave of the Winds gift shop, I am astounded. Lit by the brilliant light of my electric lamp, dozens of small creatures crawl on the cavern wall next to me.

These tiny bugs have surely been present during each of my previous journeys into Thieves Canyon, a formerly commercial but now closed section of the famous visitor attraction. But until Dr. David Hubbard pointed the little beasts out to me, I had been ignorant of their existence.

David, a Virginia biospeleologist, had accompanied caver Dan Sullivan and me on a trip into lower Cave of the Winds to find and identify previously

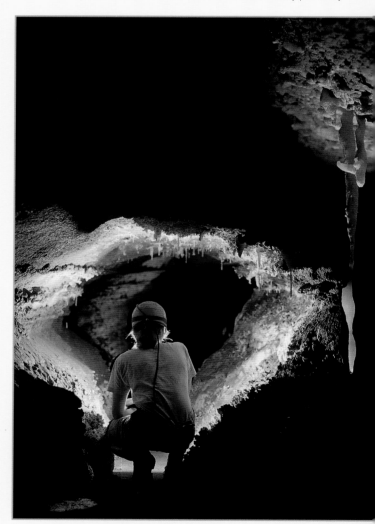

Steve Lester illuminates a passage and formations in Breezeway Cave's Happy Trails; an array of living creatures makes their home in such passages.

unknown cave life. Dan and I watch while David excitedly scoops up one specimen after another. In a way, it seems cruel to snatch away these unwilling participants for their role in the advancement of science, but their capture and study will help biologists learn more about the animals that live in Colorado's high caves.

Thieves Canyon is a particularly rich site for collection. Old pieces of wood from rotting ladders and long-extinguished lighting systems as well as abundant moisture that turns the floor into squishy mud both provide nourishment for these subterranean dwellers. The bugs are everywhere: on the walls, floor, and probably the ceiling. Though I know they're tiny and harmless, I still look for a rock containing the least amount of life on which to set my cave pack.

The collected specimens from our evening in Thieves Canyon prove interesting for David and his colleagues. Prior to his 1996 Colorado visit, biological collection in the state's caves was extremely limited. A. S. Packard visited Huccacove Cave in lower Williams Canyon in 1888, collecting and identifying a beetle and two flies. In 1961, biologists visiting Fly Cave near Cañon City found several interesting specimens, including a new spider species. In 1980, Dr. John Holsinger collected insect specimens from Fulford and Blue Butt caves near Eagle.

On our trip into lower Cave of the Winds, David explains that biological study of cave life, particularly in the American West, is remarkably incomplete. Almost any cave might harbor a new species of life unknown to scientists. His collection of life in Thieves Canyon includes several possible new species and a variety of undetermined millipedes, mites, springtails, and beetles.

David visited many other Colorado caves during his trip. In Manitou, Fly, Marble, Groaning, and Spring caves, David collected numerous specimens; many of these species were previously unknown.

Bats live in many Colorado caves, consuming thousands of insects in a single night during summer.
PHOTO BY WENDY SHATTIL AND BOB ROZINSKI.

Caves hold a huge amount of life, from animals that visit the caves on a temporary basis—such as bats, cave mice, spiders, and flies—to those that spend their entire lives underground either in the twilight zone or in total darkness. Though Colorado has no blind underground fish, on occasion, trout washed in from surface streams have been found stranded in caves. Cave mice are common. In Glenwood Caverns, cavers sleeping overnight in the Bedroom chamber off the commercial tour route must carefully seal their food in containers to avoid feeding the hungry mice. One caver was once startled awake by the feeling of little feet crawling up her outstretched arm.

Bats frequent many Colorado caves, usually living in entrance areas where the temperature is consistent. A huge colony of Mexican freetail bats spends summers in the abandoned Orient Mine north of the Great Sand Dunes National Monument, the northernmost colony in the United States. At Groaning Cave, biologists observed bats patiently standing in line to travel through a small bat crawlway past a cemented fissure near the entrance. To assist the bats, the Forest Service removed part of the cement blockage and installed bat-friendly bars that allow the mammals an easier route in and out of the cave. On a trip deep into Fixin'-to-Die Cave, I once was amused by a bat that landed on the wall of a fissure near me. Strolling up, it seemed curious as to why I was so deep in the cave. I wondered the same, as this was a section that had only been recently opened through a long and difficult dig.

James Meacham, a bearded Colorado Springs biologist, spent several years in the early 1980s studying the lampshade spiders found in the caves of Williams Canyon. Originally identified in Fly Cave near Cañon City, this rare, harmless spider, which builds lampshade-shaped webs in which to trap its prey, is known only in a handful of caves along Colorado's Front Range. In two decades of exploring the caves of Williams Canyon, cavers have noticed that the spiders will move away when a cave becomes too frequented by humans.

In the caves of Carlsbad Caverns National Park in southern New Mexico, microbiologists have begun to seek out microscopic life. They've not gone away disappointed, as new and unusual microbial life is found on every expedition. Some of these life-forms are exceptionally bizarre, as caves may hold life unchanged over thousands of millennia—true living links back to an earlier Earth. Though no one has yet sought out microbial life in Colorado caves, it is almost certainly here.

Ignorant of the life underground, humans can destroy entire species or cave ecosystems through pollution or the introduction of other opportunistic life. In one deep New Mexico cave, biologists are studying the damage from the accidental introduction of common coliform bacteria, brought in and spread throughout a chamber by humans. Though present in all mammals, these bacteria invade the underground environment and might eventually destroy native microbial life.

Collecting, identifying, and cataloging just the visible life in caves could take hundreds of years. David Hubbard reports that very few biologists today have the time or the interest to collect and identify. No one knows how many species living in the caves of the American West are still awaiting identification. If David's one-week trip to various caves of Colorado is any indication, there could be hundreds, if not thousands.

Marked with a whimsical sign for Halloween, Snider's Cave above Cave of the Winds likely predates many of the caves in the Williams Canyon/Cavern Gulch area because of its high elevation and formation by the mixing waters process.

On the central White River Plateau, quartzite rock holds an unusual series of caves. Normally nonsolutional, the calcite cement in this rock was dissolved by draining glacial water along cracks and crevices. The resulting caves measure up to 600 feet or more in length.

OTHER GEOLOGIC PROCESSES

In Colorado's arid and semiarid regions, caves have been found in gypsum outcroppings. Far less sturdy than caves of limestone or dolomite, gypsum caves are frightening places to visit. In the early 1990s, a quarrying operation along the Arkansas River near Howard revealed a small gypsum cave. Like similar gypsum caves near Eagle, this discovery attracted cavers, who visited with permission. Once inside, however, the cavers quickly turned around, heading for the safety of the surface. To their horror, they discovered the cave's ceiling and walls contained many large cracks—at any moment, the cave could collapse.

Near Grand Junction, draining water eroded surprising caves in dried mud. Along badland gullies, Dr. Ronald Delano and Donald G. Davis discovered several caves, some hundreds of feet in length. Often multi-entranced, these mud caves appear very similar to limestone solutional caves and are much more stable than gypsum caves. The infrequent run-off of snowmelt and rain forms these caves through the erosion of the clay over very long periods of time. Near Rulison, Anvil Points Claystone Cave is the longest mud cave known in the world, with greater than 2,000 feet of surveyed passage.

Colorado is host to several noteworthy granite caves. Although granite talus caves are common, the caves on the slopes of Pikes Peak and those of Lost Creek near Deckers are unusual in their geologic development and length.

With a depth of 553 feet, an obscure Pikes Peak granite cave is currently the world's deepest of this type. Discovered in 1991 by Colorado Springs caver Mike Frazier, Hurricane Cave has been surveyed to over a half-mile in length. Geologists report that Hurricane was formed following a fault along which a stream now flows. This stream continues to slowly erode the granite bedrock and boulders covering the stream channel. Decomposition of the granite allows the stream to carry away the resulting sand, enlarging the cave. Eroding granite on the surface, along with fallen leaves and other organic material, forms soil above the cave, allowing flora to grow there.

The many caves within the Lost Creek Wilderness are developed in a similar manner. Streams flowing along joints or crevices in the granite slowly erode the rock, while boulders above the watercourse erode through chemical decomposition. Collapsing inward, these rocks form a roof to the cave below. Other western states have similar granite caves, but Colorado has perhaps the longest and best developed.

Stocking the Store of Knowledge

Reaching the soggy University of Colorado campsite, Mark, Paul, and I load our caving packs in a brief moment of sunshine. Other members of our group arrived the previous evening and tell us of the steady rain that lasted through the morning. We take off on foot as the skies darken again, the smell of rain blowing in from the west.

Here at the Colorado–New Mexico border, we learn that the entrance to the lava tube we are to survey was completely excavated over several decades. The lava tube's discoverer claimed to have found it through his dreams, as it has no open entrance. I wonder if other lava tubes in the region, particularly north of the border, will also need excavation to enter. I can see in the distance a huge wooden windlass standing guard over the shaft at the lava tube's entrance, like a relic from an ancient civilization.

Geologists and scientists have helped make the discovery of new caves easier. As prospectors found most of the obvious Colorado caves in the late 19th century, current cavers rely more on geologic knowledge to find caves today. In 1974, Mark published a helpful guide to the White River Plateau, identifying trends and tendencies of the caves of the region. Two decades later, Paul wrote a master's thesis at the Colorado School of Mines on the high-altitude caves and underground drainage of the Lime Creek region south of Eagle.

Hydrological connections between sinkholes, caves, and springs are well known and studied in the eastern United States. In Colorado, less is known of such hydrology, as only a few studies have been conducted. Paul's thesis and Fred's study of the Williams Canyon caves are probably the most detailed to date, establishing relationships between sinkholes that sometimes lie many miles away from the springs.

During the 20th century, geologists and town boosters believed the water emerging from the mineral springs of Manitou was thousands of years old and unaffected by drought, pumping, or pollution. Though this seemed ideal, this belief of virgin water was unsubstantiated.

Geologist John Thrailkill dye traced Fulford Cave's stream in 1959. The bright green water appeared in the spring at a nearby beaver pond some 15 hours later. PHOTO BY JOHN STREICH, COURTESY JOHN STREICH COLORADO GROTTO COLLECTION.

In his studies of the canyon's hydrology, Fred suspected the springs were fed not by an ancient underground reservoir, but by area streams and water seeping into the ground. In early June 1998, Fred tested his theory by dumping fluorescein dye into the Williams Canyon stream above where the water annually sinks into the streambed, seemingly never to reappear. Placing a collection device in a small mineral spring along Fountain Creek, Fred began his wait for the appearance of the dye to confirm his theory. Throughout the summer, he checked and replaced the device, finding no trace of the dye. As summer turned to fall, he began to wonder if his theory was somehow wrong.

By the end of the year, with no dye visible in any of his collected samples, Fred chose to examine the samples more closely. Following instructions from other hydrologists, he found that the dye had actually reappeared unnoticed in the mineral spring in early September—about three months after the start of the test. Collection devices placed in the spring several months afterward continued to show minuscule traces of the dye, diluted by groundwater.

For Manitou Springs, this positive trace indicates that the springs are not indestructible. They are susceptible to drought and can be polluted by any of the water sources entering the system from either Ute Pass or Williams Canyon. Fortunately, water entering the reservoir is fairly clean. Plus, the limestone rock and network of caves that form the reservoir act as a natural filtration system.

Increasingly, cavers apply knowledge from geological studies and previous exploration in known caves to identify likely locations at which to dig for hidden entrances. Sometimes, something as subtle as a different type of dirt or moss along a crevice can serve as a clue to an undiscovered cave.

During the 1993 excavation of the Breezeway Cave entrance in Williams Canyon, wind that blew from an inconsequential hole behind a large rock was a clear sign of hidden passage. Digging in, our group removed a large amount of dry and dusty guano, indicating that packrats had long lived in the windy crawl.

In Breezeway Cave's Elkhorn Chambers, helictites fragile enough to break at the lightest touch reach out to Bill Allen.

Fortunately, Colorado caves seldom conceal dangerous or unpleasant life other than the occasional rattlesnake in a low altitude cave. In Williams Canyon, a cautious caver once came upon a hibernating black bear in a Piñon View Cave crawlway. At Hubbard's Cave, cavers spotted a snowshoe rabbit in the Eastern Parallel during several trips. In the White River Plateau's Doublesink Cave, cavers discovered a trout that had been washed into a pool during high spring run-off. Unable to swim back outside, the trapped trout was captured and served as dinner. Other caves are home to swallows, owls, and bats.

In warmer climates, other species make caves their home. Salamanders, crickets, and harvestmen live in New Mexico's caves. In Colorado, the high altitude and cold interior temperatures make it difficult for most animals to survive (see "Who Lives Underground?" on page 60).

Reaching the windlass, we had no fears of rattlesnakes that might be sunning themselves on the rocks piled alongside the lava tube's entrance. A steady rain falls upon our group, now completely obscuring the mountains to the west. The tube's entrance is surprising in its depth and extent; a rickety wooden ladder held in place by a rusted wire provides our admission to the cave below. Squeezing through strands of barbed wire surrounding the shaft, we climb down the shifting ladder to the tube's dirt floor.

Here, out of the weather, we begin our survey. Paul's survey team takes the larger southern branch of the tube, while my team accepts the shorter northern branch. Though our branch is the shorter of the two, we're happy and dry. Although science had nothing to do with the discovery of this tube, our survey adds to the knowledge of the underlying basalt flows. Perhaps—someday—our survey will lead to the discovery of Colorado's first lava tube.

OPPOSITE: The astonishingly delicate and rare beaded helictites in Cave of the Winds' Silent Splendor rank among Colorado's cave wonders.

Old Bones and Cavemen

During the 1982 Coke Bottle Crawl dig adjacent to the Cave of the Winds commercial trail, tour groups passing by our troop of muddy cavers often were surprised and amused. Usually, they would stop and talk.

"What are you looking for?" they'd ask. "Gold? Bones?"

We'd tell them "nothing," which wasn't far from the truth—breaking into a new passage results in finding only air. The joke always brought a laugh, but in reality the visitors weren't too off-base when they wondered if we might uncover bones in the fill.

Throughout Colorado, bones have often been found in caves. In Cave of the Winds, cavers digging in the Cliffhanger section in 1986 discovered a buried toe bone of an extinct Pleistocene-era horse. Identified by a University of Colorado paleontologist, the bone is reported to be at least 12,000 years old and possibly more than a million.

Another group of cavers digging in 1993 in a small cave on the White River Plateau east of Glenwood Springs unearthed several large bones. Believing they were from a cow, the cavers piled the bones elsewhere in

Jon Barker, Chas Lindsey, and Wesley Cronk explore Narrows Cave, where Pleistocene horse bones and a 250,000-year-old ground sloth scapula were discovered; Narrows is one of three places in Colorado where remains of the extinct sloth have been found. PHOTO BY CAROLYN ENGLUND CRONK.

Colorado's caves can conceal the bones of ancient animals such as this ringtail cat—as well as human bones.

the chamber and continued their excavation. To their great surprise, they soon exposed the skull of a buffalo.

Realizing the significance of their find, the cavers carried the skull to the Denver Museum of Nature & Science, where it was identified as belonging to an extinct mountain buffalo. Further study by the museum indicates the collected bones are well preserved and are a nearly complete skeleton—one of only a handful discovered in Colorado's high country.

Across the state, a seldom-visited tectonic cave (a natural cavity formed by the faulting action of the rock) in the mountains near Saguache has an unusual feature. To the astonishment of Colorado Springs caver Walt Rubeck and the cavers accompanying him, they found within this cave three bighorn sheep skulls. Curiously, they were piled together, almost as if someone had deliberately stored them in the cave.

During the summer of 1978, National Park Service personnel from Lincoln, Nebraska, partially excavated a strange volcanic bubble cave near Gunnison. Called Haystack Cave, this short cave has a rich collection of Pleistocene vertebrate remains. Carbon dating of some of the bones indicates approximate dates of 12,000 to 15,000 years ago.

Colorado's most significant cave for Pleistocene bones is the privately owned Porcupine Cave. Accessed through an old timbered mine tunnel, the cave had been explored by cavers for many years before Don Rasmussen came upon some bones during a 1981 trip. In 1985, Pittsburgh's Carnegie Museum of Natural History began preliminary excavation in the cave, and the dig has continued most summers since then.

Currently managed by the Denver Museum of Nature & Science, excavations in several locations within Porcupine have established it as one of the richest Pleistocene bone storehouses known in North America. Scientists have found more than 100 species during their digs, including bones of llamas, camels, cheetahs, horses, ground sloths, wolves, foxes, black bears, and skunks. The bones date from between 300,000 and 2 million years before present—a period when the climate of Colorado changed numerous times, clearly noted by distinct layers of fossils.

The 1986 discoveries of several previously unknown chambers by Don and fellow caver Kirk Branson were key to the identification of several of the prime sites within Porcupine Cave. These new rooms, having been sealed off

Cavers hike near the privately owned Porcupine Cave, home to one of North America's richest lodes of Pleistocene bones.

from the remainder of the cave, were pristine and free from visitor disturbance.

Throughout the excavation, National Speleological Society cavers have participated on several occasions. Cavers have helped carry bags of dirt to the surface for careful screening of small bones and, in 1997, resurveyed Porcupine to create a detailed map for use in scientific papers. Currently, Porcupine has just over 2,000 feet of known passage.

Another one of Colorado's caves gained national attention for a July 1988 discovery. In Hourglass Cave in the White River National Forest near Vail, cavers Rich Wolfert, Cyndi Mosch, and Tom Shirrell came upon a disquieting find—human bones in a passage thought previously untraveled. Recognizing that their find was exceptional, the following summer they brought in caver and anthropologist Dr. Patty Jo Watson of Washington University in St. Louis for further study and interpretation of the bones.

Patty Jo believed the bones were very old, predating recorded history. The bones were determined to be from a 35-year-old male who died 8,000 years ago—one of the oldest such skeletons yet found at high altitude in the country. Upon closer examination, cavers found numerous charcoal fragments on the floor and smudge marks on the cave walls leading to the skeleton. Quite likely, early explorers used pine torches to travel deep inside Hourglass Cave for reasons as yet unknown.

In 1993, following nearly 5,000 hours of study, the bones were presented to the Southern Ute Tribe for burial. Tribe representative Kenny Frost believes the bones belonged to a powerful spiritual person who traveled into Hourglass Cave to die in peace. Others believe the explorer might have become lost, disoriented, or injured in the cave. His pine torch might even have extinguished accidentally, leaving him alone in the dark, unable to find the route out of the complicated maze.

Hourglass Cave is still a well-kept secret today. Gated by the Forest Service, the cave also contains numerous bones of area animals, including black bear, bighorn sheep, deer, fox, and marmot. It is open only to qualified scientists for future study of the history of this unique Colorado cave.

Pearls on a String

Deep within Cave of the Winds, we pick at a mud plug in the passage. Located only a few feet from the Old Curiosity Shop, the low-ceilinged, mud-filled crawlway was discovered when work crews excavated the main passage for a new commercial tour route.

Taking advantage of an opportunity to excavate a passage leading away from the known cave, four of us work with determination. Lying on my belly, my arms outstretched, I dig at the face. Behind me stand Donald G. Davis, Mark Maslyn, and Jim Barber, who take the mud I pass behind for disposal. Abruptly, the mud crumbles, revealing a hidden cavity beyond. Peering into the newly found hole, I grin and back out.

"Donald, take a look at this," I suggest.

Donald climbs past me and over an electric light into the muddy crawl. A moment of silence follows, then, "Oh, my!"

In our effort to find an extension leading to a new cave, we instead discover a small cavity not more than six by three feet. What makes this cavity so unusual are the beaded helictites that fill it like the web of a spider. Beaded helictites are an infrequent speleothem, or an uncommon mineral deposit, known only in a handful of caves worldwide. For unknown geologic and mineralogical reasons, Williams Canyon contains two caves with profuse displays of the rare formation.

During the winter of 1983–1984, an extensive dig project off of the Breakdown Room in Cave of the Winds revealed the previously unknown Silent Splendor, a 200-foot-long corridor containing exceptional displays of white beaded helictites clustered along its western wall. This treasured chamber and its helictites were later the subject of a Denver public television station documentary.

Beaded helictites are exceptionally delicate, like pearls on a string. Defying gravity and growing in any direction, this naturally occurring mineral formation can extend for two feet or more in length. Literally covering some walls and pockets, these crystalline displays consist of aragonite, a variant form of calcite. Beads can range from as small as a pinhead to as large as marbles.

At Cave of the Winds, developers discovered the uncommon formations in the mid-1880s when they opened for visitors passages not far from our 1989 crawlway dig. Paying customers were thrilled to behold the dazzling display of white beaded helictites covering the ceiling of the famous Crystal Palace. Unfortunately, the dry air of the tourist route and the close proximity to visitors doomed this beautiful sight. Within only a few decades, this once-spectacular chamber was dry and lifeless, with large pieces of the helictites broken off by thoughtless visitors.

While Silent Splendor is located some distance off the commercial tour route and probably will never be publicly shown, our new discovery is only 15 feet off the route. In discussions with the cave's management, we decide to carefully cover over our Spider Web Valley for some future time when technology will allow the cautious showing of the slender helictites yet protect them from visitors.

Though geologists have studied these unusual mineral formations, a consensus on how they grow has yet to emerge. Most believe all helictites are formed through slow capillary water dripping from the base. Water flows through tiny apertures in the structure of the helictite so slowly that mineral deposits build up. Water supply, airflow, evaporation, impurities in the water, and hydrostatic pressure help the helictites grow in defiance of gravity. Yet it remains a mystery why some should grow in tubelike forms while others adjacent to them grow in beads. No one can guess how long it takes a helictite or even a single bead to form.

One clue to the growth of beaded helictites is that all examples found in Williams Canyon caves are located along the up-dip side of the passage, the area where the rock layers slant upward. Calcite-rich water moving down-dip along the limestone beds apparently emerges from the rock into these unusual formations. In Silent Splendor, the up-dip wall has numerous groups of beaded helictites, while the passageway's down-dip wall is barren.

Despite the rarity of these formations, Breezeway Cave near Cave of the Winds has probably more known beaded helictites than any other cave worldwide. Chamber after chamber is filled, in areas both large and small. Heaven's Gate is the cave's showplace, with an entire wall brimming with thousands of helictites growing over and on top of each other.

Curiously, numerous broken helictites pepper Breezeway's floor. Some geologists believe these are the result of helictites growing too long to support their weight; others believe the helictites were severed by shock-waves from blasting at a nearby limestone quarry.

Whatever their origin, beaded helictites are as fascinating as they are beautiful. Displays such as Silent Splendor's Sea Anemone are now known around the world. In the late 1980s, one of the most acclaimed European cave photographers was puzzled when his caver guides brought him to Cave of the Winds. En route from the caves of South Dakota's Black Hills to Carlsbad Caverns, the group made their way from the end of Cave of the Winds' commercial tour route through the Breakdown Room and Whale's Belly to Silent Splendor. Upon spotting the Sea Anemone, he understood the reason for the detour. Pulling out his expensive photography equipment, he was overjoyed at having the opportunity to see and photograph one of his favorite formations. ⩗

OPPOSITE: Cavers call this basketball-sized group of beaded helictites the Tesla Coil. It is found in Breezeway Cave's Elkhorn Chambers.

Adventures
in Surveying

Despite my best efforts, my survey book sketch bears little resemblance to the actual passage in which I stand. The cave is simply too complex. Pulling the metal cap off my mechanical pencil eraser, I delete a good portion of what I have just drawn. This sketch is going to take some time.

In the distance, I can hear the hoots and hollers of my companions, cavers Al Hinman and Paul Fowler. Each is calling to come see their discoveries.

Shaking my head, I set aside my survey book on a nice, flat rock and slip the pencil underneath the rubber band. The survey can wait. I want to see some of this new passage.

The three of us are two hours into Groaning Cave, Colorado's longest known cave. Though it is Fourth of July weekend, and at least three other groups are in the cave, we haven't seen anyone for hours. Of course, with 10 miles of surveyed passage, you can lose an entire marching band in Groaning and never see them again.

I led our little survey team back into the G Survey on behalf of Alan Williams, the Lakewood caver who is coordinating the continuing survey. Though Alan has promised a new set of quadrangles of the cave for nearly a decade, we don't expect to see it anytime soon. For now, Dr. Norman Pace's 1975 pocket map and the out-of-print 1973 wall map stretching eight feet long are the best maps available.

Dropping below the main level of the passage between some large rocks, I observe that the low crawlway before me appears to be virgin passage. Although this particular crawlway is less than two minutes from the cave's main trade route, which hundreds of cavers have traversed, no one has ever attempted to crawl down it to see where it leads.

Deciding that now is as good a time as any, I squeeze in, my helmet bumping the ceiling. Loose, crumbly gypsum sparkles in my headlamp's light and showers me as I push and pull my way through the crawl. The passage opens a little, allowing me to rise to my hands and knees. Ahead, I see it becomes impossibly tight: so much for a big discovery.

Returning to my book, I find Paul waiting. He tells me of the passages he has explored. They remind him of the multiple levels of South Dakota's Wind Cave, where surveyors have taped, or measured, more than 95 miles of passage. Al is off exploring some lower-level passageways, Paul reports.

Groaning is an exceptionally complex cave with a maze of passages on at least three levels. No one knows how many miles of passage might eventually be found and surveyed in the high-altitude cave; estimates range from 12 to as many as 20 miles. Surveying has played an important role not only in the cave's exploration, but also in defining the cave's limits.

When I first began caving in 1974, Groaning seemed to have no boundaries. Its length seemed incredible as I rolled out Norm Pace's cave map on the floor at my first Colorado

ABOVE: Alan Williams makes his way along a sculpted passage deep within Groaning Cave, the state's longest known cave.
OPPOSITE: June Harris plots a route into the east entrance of Hubbard's Cave.

Peter Jones uses a Brunton compass in a 1971 survey of Groaning Cave. PHOTO BY NORM PACE.

Surveying Basics

Although explorers once declined to participate in surveying efforts, they now willingly assist. Not only do they usually get to choose where the survey stations are placed, they also have plenty of free time to explore the passageways around and ahead of the survey.

Surveys move at the speed of the sketcher. Some sketchers pride themselves in being speedy. Others, like me, take their time to make certain the sketch is accurate and representative of the passage. Floor detail is usually a requirement in most surveys these days, as is a continuing profile of the ceiling of the passage. Some surveys even collect information about the geological features of the cave through which they pass.

The development of powerful and inexpensive personal computers since the late 1980s has greatly helped surveying. In the pre-computer days, survey notes and sketches would be laboriously plotted by hand, a task taking weeks or months for larger caves. Computers allow data to be entered into graphic plotting programs that reduce the data, plot the cave, and even provide profile views that can be rotated 360 degrees in real time.

One of the more popular plotting programs is Compass, a Windows-format software program developed by Denver caver Larry Fish. Designed specifically to help plot mountains of data for Groaning Cave, the program is now used by cavers throughout the country, as well as by many units of the National Park Service. Cave specialists at both Carlsbad Caverns and Wind Cave national parks use it to plot their caves.

Grotto meeting. Because Groaning had only been known at that time for five years, anything seemed possible. Norm pointed out the passages shown on the map: "These are the main trade routes," he explained. "There are many passages that haven't been surveyed yet."

I imagined Groaning could expand in all directions, with a multitude of rooms and corridors yet to be found, much less surveyed. I eagerly looked forward to seeing the cave for myself and contributing to the knowledge of its extent.

What Norm didn't tell me was that the cave had reached its boundaries on both sides and along the back. Faults isolate the limestone block in which Groaning is developed, and as far as we know, there is no way through the fault zones, even though geologists suspect there is more cave on the other side.

Surveying helped identify these boundaries to the cave. On a map, the faults are obvious, cutting a straight path along the landscape, terminating big and small passageways alike. In the 20 years between my first glimpse of the Groaning map and my survey trip with Paul and Al, maps of the cave still had the same general outline. The main difference is that cavers have surveyed a good number of the passages in between the trade routes.

Geologist John Thrailkill uses a Brunton compass to determine the direction of a passage in Fairy Cave, circa 1955. PHOTO BY JOHN STREICH, COURTESY JOHN STREICH COLORADO GROTTO COLLECTION.

Although cavers are turning more often to software drawing programs to draft maps electronically (Alan Williams is using AutoCAD for Groaning Cave), the most aesthetically pleasing maps are still drawn by hand. Even so, the artist will usually prepare a computer line plot of the cave to serve as the backbone of the hand-drawn map. With the line plot and the original sketch from the cave, an artist can graphically re-create the cave on paper.

Sometimes, in the case of complex caves like Groaning or Narrows Cave in Williams Canyon, multiple levels can cause confusion when one tries to draw passages on top of each other. A common solution is to offset one passage or to use dashed lines to indicate a passage above or below. On occasion, maps will use clear transparencies to show overlapping areas—the viewer can "peel back" levels to see the one of interest.

Standing perplexed in Groaning's GD Survey, I can understand why field sketches can be overly simple. The complexity of caves can be overwhelming, particularly if you are the one trying to render a three-dimensional environment on a two-dimensional page. It takes an exceptionally patient sketcher to properly draw a cave.

Golden caver Paul Burger is one of Colorado's better sketchers, and his maps have won National Speleological Society awards. As surveying director for the society's Williams Canyon Project, he has overseen the survey and drafting of the maps of the canyon's many caves, including Cave of the Winds, Breezeway, Narrows, and Huccacove. Applying principles and techniques he learned on surveys of some of North America's longest caves, Paul is demanding of the sketchers working on project surveys.

Early Surveys

The caves of Williams Canyon have been survey subjects for more than a century. Colorado College students completed the first known Colorado cave survey of the newly commercialized Mammoth Cave in 1875. Unfortunately, this survey has long been lost.

Manitou Grand Caverns was surveyed next, in 1887. Horace Hovey, a national cave expert, produced an article on Manitou's caves for the respected journal *Scientific American*. With this article is a remarkably accurate survey of Grand Caverns by J. Robinson. Apparently Robinson was not allowed to survey nearby Cave of the Winds; Hovey's article includes only a crude sketch map by his son, based on information provided by George W. Snider.

In December 1909, the Cave of the Winds management hired engineer E. A. Sawyer to survey their tour route and a small cave between that cave and Grand Caverns. Discovered in 1893, this middle cave contained large and beautiful chambers worthy of development. Although Sawyer's map showed the relationship between the two caves, it was another 20 years before Ben Snider made the physical connection through excavation of a dirt- and rock-filled passage.

Surveying can be remarkably helpful in connecting caves. In February 1986, cavers discovered Cliffhanger Cave along the cliff northeast of Cave of the Winds. Because one could reach Cliffhanger only by an exposed, overhanging rappel, we were intrigued with the possibility of connecting it to the larger cave that was less than 100 feet away by our best estimate.

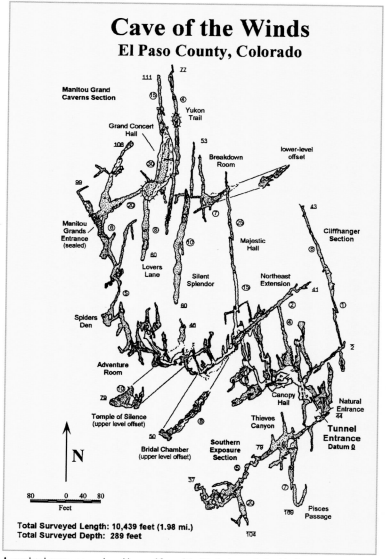

A completed survey map such as this one of Cave of the Winds shows relationships between passages as well as features within the cave. MAP BY PAUL BURGER.

Once Cave of the Winds general manager Grant Carey granted permission, Ken Kreager, Jerry Atkinson, and Steve Dunn dropped into Cliffhanger to survey. That evening, after the team climbed the rope up from the cave, they reported that they had surveyed Cliffhanger as far as they could go without digging.

From the top of the cliff, they ran a survey line along the canyon side to the Cave of the Winds gift shop, where they linked to the master survey of the cave system. Within a few days, we learned the caves were closer than we had dared to expect; at their closest points, they were less than 60 feet apart.

Armed with the new information, a large group of cavers soon returned. Digging extended the cave another 20 feet toward Cave of the Winds. At that time, an understanding Denver caver placed a perfumed rag in a possible connection route from Cave of the Winds. The wind blowing past the rag quickly carried the scent into Cliffhanger, alerting the digging team of the potential connection. Surprisingly, the connection was not made through that passage, but by a closer route indicated in the surveys.

Even though some surveys, like Groaning's, can take years or even decades, surveys are relatively quick projects. Depending on the speed and confidence of the sketcher, teams in Colorado can survey 500 to 1,000 feet in a single day. Because most Colorado caves are less than a mile in length, surveys usually are completed in a few weekends or a summer at most.

Horace Hovey's 1887 map of Manitou Grand Caverns remains remarkably accurate even today; the numbers refer to rooms, passageways, and features.

To determine Cliffhanger Cave's relationship to nearby Cave of the Winds, cavers surveyed along high cliffs to the Cave of the Winds gift shop (at right) in the 1980s.

Why Survey?

During the 1990s, one of the state's most prolific surveying teams was Hazel Barton, a microbiologist from Great Britain, and Evan Anderson, a Denver musician. With Paul Burger directing survey efforts in Williams Canyon, Hazel and Evan turned to the rest of Colorado. They found that many caves needed to be surveyed or resurveyed, as most had last been actively surveyed in the 1950s or early 1960s by Colorado Grotto members.

In short order, Hazel, Evan, and several teams of cavers resurveyed the Clear Creek Fault Cave west of Golden, South Park's Porcupine Cave, Cave Creek Cavern near Fairplay, and Glenwood Canyon's Hubbard's Cave. For every cave, their efforts produced more complete, more detailed maps than the initial surveys.

Their biggest challenge, though, was the resurvey of Fairy Cave. Aware of the team's continuing survey projects across Colorado, Glenwood Caverns manager Steve Beckley specifically requested that Hazel coordinate a new survey to indicate the relationship between the cave's upper and lower levels. Fairy Cave had only been partially surveyed during a Colorado Grotto project that ended in 1968. Many of the passages Steve had explored in his trips were unsurveyed and not indicated on the map provided by Pete Prebble, the cave's owner since 1961.

Beginning in late September 1998, Hazel and Evan launched a several-year project to resurvey the entire cave and survey any new passages discovered by members of the Fairy Cave Project. They started with the historic commercial trail on the cave's upper level. Unlike the survey of the commercial Cave of the Winds, during which surveyors discovered that the steel railings affected the compass readings, the Glenwood Caverns trail survey was achieved with ease.

Survey teams quickly moved to interesting undeveloped passageways leading to the cave's lower levels. Digs at four separate locations promptly revealed new passages and rooms, adding several thousand feet of passage to the cave. At the end of the survey project's first year, Glenwood Caverns boasted two miles of surveyed passage; the 1960s Colorado Grotto survey had less than 8,000 feet of passage altogether. Already, the new commercial cave had gained in stature.

Computer survey plots of Glenwood Caverns fascinate geologists. The cave's profile shows the historic tour route is nearly level, indicating an ancient stream terrace where the prehistoric Colorado River flowed for millennia. Below this level, the cave drops steeply, a sign that the Colorado River aggressively eroded a deeper channel owing to changing climatic or geologic conditions, forming Glenwood Canyon. Surveys of the nearly level Hubbard's and Groaning caves tell geologists they formed under geologic conditions in which the water table remained stable for millions of years.

For cavers seeking undiscovered passage, surveying can show areas where they might find new passage. On the Groaning Cave map, a large blank area to the south of Serenity Hall fell well within the confines of the boundary fault yet had no known passages. On my first trip to the cave in 1974, Golden caver Rich Wolfert and I spent hours poking and pushing tight holes and fissures along Serenity in an attempt to find the theorized passage to the south. Though we were rewarded with some nicely decorated chambers, none of the passageways continued for any distance.

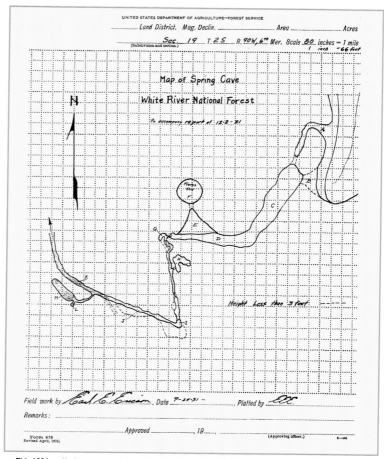

This 1931 preliminary map of Spring Cave by White River National Forest ranger Earl E. Ericson is the first known map of the cave.

A Survey Day

High above the Williams Canyon floor across from Cave of the Winds is a small cave seldom visited. With less than 100 feet of passage, two entrances, and a grand view of the canyon, Easter Dome Cave contains little more than its impressive namesake ceiling dome.

In a canyon filled with small caves, Easter Dome is important to geologists, who consider it proof that caves once crossed the width of the canyon. Clearly a remnant of a longer cave now lost to down-cutting, or eroding, by the Williams Canyon stream, Easter Dome also contains an unusual deposit of ash in a short side passage. Quite possibly, the ash is the remains of an ancient cave rat midden that once burned, perhaps ignited by a lightning strike or forest fire.

Because of Easter Dome's significance, members of the Williams Canyon Project decided it should be surveyed. In addition, it needed to be tied into a growing overland survey that will eventually link all significant canyon caves to an accurate grid. With such a master map, relationships between caves, as well as possible extensions, will be more easily identified.

Appropriately, on Easter weekend in 1990, I led a group to this scenic cave. Climbing the steep, overgrown eastern wall of the canyon, we enjoyed the view up-canyon, toward the Rampart Range. To our surprise, another group of adventurers was scaling the western wall to the large oval opening of Octagon Cave, a single-room cave reached only by an exposed climb.

Keeping a watchful eye on the climbers, we continued our ascent to the low northern entrance of Easter Dome. At the entrance, our party split into two groups. I led one group around an easy cliffside ledge to the cave's southern entrance. As we emerged into full view of the Cave of the Winds balcony, our appearance delighted visitors on a cave tour. Ignoring their shouts, I led my companions into a low side passage where they pulled from their packs folding shovels, gardening tools, and other implements of excavation, ready to have a try at a possible extension.

Once their excavation was under way, I hurried back to the others. At the cave's overgrown northern entrance, they were fighting the brush oak and high grasses to set the survey's first station. Because I was to keep notes, I quickly pulled my survey book and pencil from my pack and awaited the first shot, or round of gathering and recording data. Jacobe Rogers handled the compass and inclinometer, peering through the instruments that hung by a cord around his neck. I sketched as the survey team entered the cave, the point man backing his way through the crawlway as he led the team to the next survey station.

Taking only a half-dozen or so shots to complete the cave, we reached its southern entrance just as the last diggers gave up in disgust on their new dig. It seemed the fine ash had taken its toll on the crew, driving caver after caver out into the sunshine. A cloud of ash billowed from the side lead, a smaller tube connected to the main passage; fortunately, it would take only two or three shots to complete our survey in this section.

Holding our breath as best we could, we quickly placed our stations and retreated to the fresh air outside. Because it was only noon, we decided that after lunch we would find the nearby Donahue's Cave and survey it, giving us two surveyed caves for the day.

Just east of Easter Dome, Donahue's Cave is an equally short cave I found in February 1974 with two high-school friends. On our first trip to Williams Canyon, we grew tired of the underprepared amateurs who crowded popular Huccacove Cave down-canyon. Until Cave of the Winds securely gated the cave in 1978, partygoers and adventure-seeking youth who swilled beer and liquor around smoky campfires often chose Huccacove as their destination. Appalled at the misuse of the cave, we three followed the June 1880 suggestion of the Reverend Roselle T. Cross,

Octagon Cave overlooks Williams Canyon from high on a cliff face.

who led his boys' exploring club away from Huccacove. Outraged at an admittance fee of 50 cents per boy to the fading commercial attraction, Rev. Cross declared they would find their own. They did just that: The young Pickett brothers pushed through a low squeeze to rediscover the inner Cave of the Winds.

Nearly a century later, Chris Bakwin, Brian Donahue, and I had similar luck. After leaving Huccacove, we visited Easter Dome and contoured around the mountain in search of other caves. To our surprise, we found a hidden entrance, and named it after Brian, who first spotted it in the heavy brush. Leading south, the cave's single passage was blocked by dirt and organic debris. A single white stalactite in the back proved visitors did not often examine it, for nearly every other cave in the canyon had been badly vandalized.

Unfortunately, by the time I led the survey party to Donahue's Cave, the stalactite was long gone—all that remained was a broken stub. Someone had dug at the floor also, revealing a slender squeeze to an inner passage. As the survey crew prepared for another survey, we sent the thinnest member of our group to the squeeze, where he clawed and pulled his way into the passage beyond. He disappeared for a few minutes, returning to confirm that the inner passage deserved to be surveyed.

As we set our first point inside the descending passage, the other cavers pushed ahead of us so they could explore. Our survey proceeded quickly, with only two shots needed to reach the squeeze. With Denver caver Randy Reck, we spent several minutes excavating the floor of the squeeze to larger dimensions so that cavers of all circumferences could enter the inner passage. I slipped through with the survey book, awaiting the rest of the team.

Surprisingly, the inner passage grew into a modest walking passage following an easy climb. Chimneying up to this upper level—climbing with my back and hands on one wall and my knees and feet on the other—I noticed very few signs of previous visitors. Some hard-to-see blue graffiti was spotted on one wall, and others pointed out that someone had attempted to dig at the back of the cave. Unfortunately, the cave ended fairly abruptly, in a tight rock pinch that even the smallest of our group could not push beyond.

Once the surveyors arrived, I returned to my duties, copying down the survey information and resuming my sketching. We finished in only a few minutes, and then passed the measuring tape up to the crowd in the upper level for the last survey shot.

Making our way back outside and up to the eastern rim of the canyon, we felt satisfied. Though it had not been a day of extraordinary discovery as we had hoped, our group had enjoyed the exploration of two small caves and the companionship of fellow cavers. Our survey efforts, too, would help provide additional understanding of the canyon and its potential new caves in the years to come.

Twelve years later, in July 1986, Steve Sims and Paul Burger pushed through a tight constriction that Denver caver Larry Fish had discovered at the end of Hard-to-Get Hall. Following a low crawl they called Elf's Delight, Steve and Paul soon found themselves in a large corridor with numerous leads. Pressed for time, the two returned to Larry Fish and fellow caver Don Doucette, who waited anxiously in Hard-to-Get Hall for their arrival. Recognizing this as probably the big discovery cavers had been seeking for many years, they made plans to soon return.

As is often the case, word of the "Lost Horizons" discovery interested many area cavers, and the rush was on the next weekend. Several teams surveyed nearly 600 feet of passage, Groaning's largest new discovery in more than a decade. The new survey fit nicely into the map of the cave, filling in an area south of Serenity that had intrigued cavers for many years.

Working on my field sketch of the GD Survey, I knew discoveries like Lost Horizons had grown increasingly rare here in Groaning. Although revelations of new chambers or passageways in areas "bracketed" by other known passages still occur, efforts to push beyond the boundary faults always meet with failure.

But the theory that more cave might exist beyond the boundary faults in adjacent limestone still fascinates me. Perhaps cavers can use the survey of Groaning to define possible areas on the surface in which to investigate more fully. Though the adjacent rock is faulted into separate blocks, the same joint system on which the known passages of Groaning developed might continue. If so, Groaning could have a companion patiently awaiting discovery.

Closing my book, I call back Paul and Al from their lead-checking activities. We have another shot to complete, which involves measuring the passage's direction, angle, and length, then recording the data in our survey book and drafting a sketch map. After that, we're off to the GZ Survey, where Alan Williams wants to field check a survey closure problem.

"On station," Paul calls, setting the next survey station.

NEXT PAGE: Hazel Barton and Dan Lins measure distance during a survey of Glenwood Caverns.
PAGE 80: Hazel Barton examines calcite stalactites in Glenwood Caverns' King's Row.
PAGE 81: Clusters of aragonite and calcite fringe a stalactite hanging in Groaning Cave.

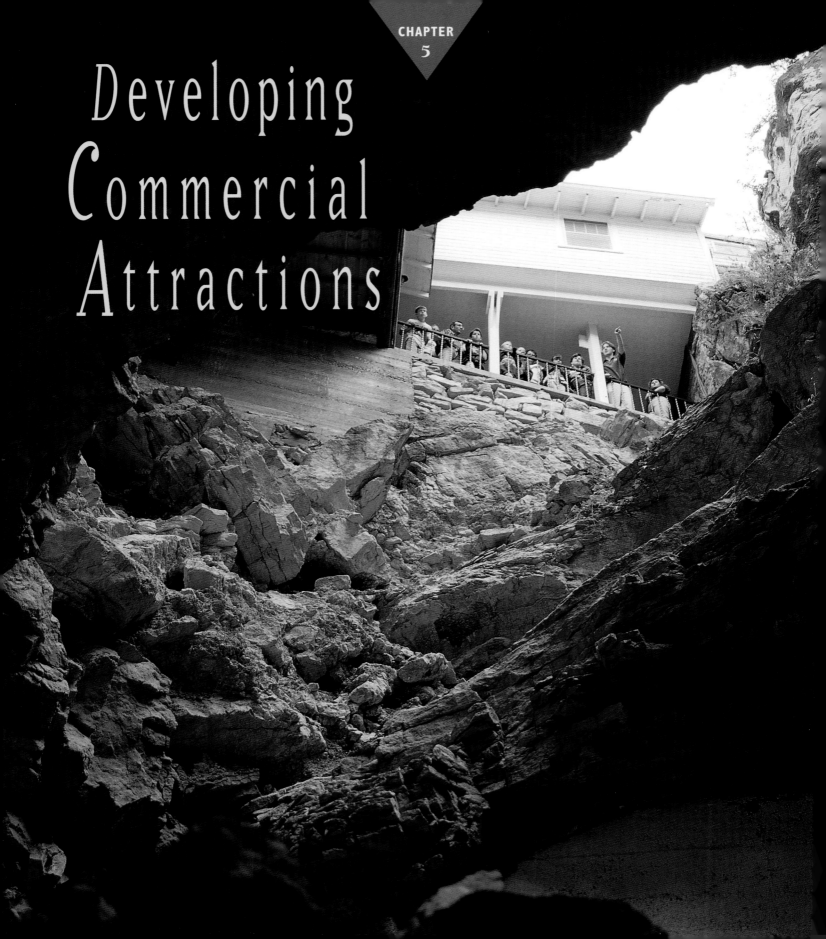

Developing
Commercial
Attractions

Closing the wooden airlock door behind me, I hear the sound of the construction ahead. Plugging my ears with the plastic ear protectors handed out earlier in the day, I pass by stacked lumber and an idle table saw to the second airlock. I turn the wooden handle, enter the chamber, and secure the door firmly.

Before me stretches The Barn, one of Colorado's largest underground chambers. Here a hive of cavers is busy transforming this rugged place into a subterranean showroom, part of the new commercial Glenwood Caverns. Several cavers jostle a heavy jackhammer, its angry roar echoing throughout the chamber. Another group with shovels and rakes spreads dirt and rock over the room's uneven floor, building a sizable platform that will serve up the initial view of The Barn for future tour groups.

No one notices my entry, or if they do, they are too busy at their tasks to acknowledge me. I climb onto a large boulder above the jackhammer crew and watch as they break apart several small rocks that block the future tour trail. Dean Mussatti, a mining engineer hired to open the new entrance tunnel and clear a pathway down to King's Row, assists the crew with a drill that has become lodged in a rock. With a practiced hand, he backs the drill out and returns it to the cavers, who eagerly attack another rock.

My view is nothing short of amazing. A few short weeks before this day, The Barn was an isolated chamber reached only by several hours of hard caving. I first visited Fairy Cave in the early 1980s, during a rare trip authorized by the cave's protective owner, Pete Prebble. That day, our group enjoyed a peaceful lunch on the pile of rocks that the work crews now loudly break apart.

Although Pete once imagined opening The Barn and the cave's other fantastic chambers to public tours, it took the ambition of a Colorado School of Mines graduate to turn these dreams into reality. Steve Beckley, a successful petroleum engineer, long had interest in exploring the historic Fairy Cave. During his college days, he met with the Colorado School of Mines Student Grotto at the school's Golden campus. Steve later spent many weekends assisting in Williams Canyon digs, helping open new rooms like Huccacove Cave's Aurora Room in February 1990.

Because of Pete's reluctance to admit cavers, only a handful of trips took place in Fairy during these years. It took Steve eight years of calls and letters to gain access to Pete's cave. Once inside, Steve was convinced the enchanting cave could be transformed into a successful commercial attraction. Throughout the 1990s, Steve discussed with Pete the possibility of buying or leasing Fairy so the public could finally appreciate its beautiful lower levels. Though Pete was at first unwilling to consider parting with his cave, in April 1998 he agreed to lease Fairy to Steve and his partners.

ABOVE: Carl Bern uses a jackhammer to remove a point of rock along the commercial tour route in Glenwood Caverns.
OPPOSITE: At Cave of the Winds, past meets present as a tour group stands on a walkway above the historic entrance.

With leasing arrangements and financing in place, Steve, his wife Jeanne, and business partner Phil Kriz began the long process of transforming Fairy Cave into the new Glenwood Caverns. Opening in May 1999, it was Colorado's first new commercial cave in more than 80 years.

Early Commercial Efforts

There are currently more than 200 commercially operated and shown caves in the United States. Colorado has only five: Cave of the Winds; the adjoining Manitou Grand Caverns, featuring kerosene lantern tours; Manitou Cave with its undeveloped caving tours; the Yampah Spa Vapor Caves, a natural sauna; and Glenwood Caverns. In recent years, few privately owned caves nationwide have been electrically lit and shown to the public. More often, state parks and other government agencies develop natural underground attractions. Development costs undoubtedly contribute to the scarcity of new caves: Arizona's fabulous Kartchner Caverns near Tucson, which opened in November 1999, cost the state about $30 million.

Entrepreneurs first began preparing trails in Kentucky, Virginia, and Tennessee caves for public tours in 1825. The development of the railroads and better transportation, along with the continuing industrial revolution, led to greater accessibility of the cave regions and increasing opportunity for individuals and families to take holidays from their work. By 1875, the year that Colorado's Mammoth Cave opened for commercial tours, resorts like Manitou Springs already welcomed visitors to its comfortable hotels and guest houses. Often staying for several weeks or even a month at a single location, visitors sought out new experiences and curiosities with which to occupy their time.

Mammoth Cave in lower Williams Canyon served as an ideal destination for visitors of the Cliff House Hotel and other Manitou Springs hostelries. The cave's proprietors furnished guides and lights for touring their attraction, located only a short walk up the scenic canyon. Though a spirited adventure, scrambling about in the dark cavern probably proved too much for most visitors. In nearby Cave of the Winds, which opened in July 1880, proprietors leveled trails and installed ladders throughout the route, giving the cave a more civilized air than Mammoth's.

Despite the easier tour route, the number of paying visitors to Cave of the Winds was still inadequate to make a

A tour group waits at the Manitou Grand Caverns entrance tunnel in 1890. Historians suspect the man in the background might be cave developer George W. Snider. Now collapsed, the tunnel served as a portal to the underground world for thousands of Victorian-era visitors. PHOTO BY WILLIAM HENRY JACKSON, COURTESY COLORADO HISTORICAL SOCIETY.

Armed with fragile kerosene lanterns, a group prepares to enter the new tunnel entrance to Cave of the Winds, circa 1895. This wooden structure was later replaced with a concrete tunnel still in use today. PHOTO COURTESY COLORADO SPRINGS PIONEERS MUSEUM.

OPPOSITE: In Glenwood Caverns' King's Row, Bill Allen admires a stalagmite beneath a veil of pristine drapery formations along the commercial tour route.

Two women pause along the commercial tour route in The Cave of the Fairies, circa 1905. For 15 years, the cave was a popular Glenwood Springs attraction. PHOTO COURTESY SCHUTTE COLLECTION, FRONTIER HISTORICAL SOCIETY.

In March 1885, Snider took the principles of sound management, an easy tour route, and extensive marketing to a new level. Snider's unveiling of Manitou Grand Caverns got good press reviews, and his new venture quickly outdrew the more established Cave of the Winds on the other side of the mountain. To the delight of his customers, the showmanship that Cave of the Winds lacked helped establish Grand Caverns as the main attraction of the Pikes Peak region.

At the new resort of Glenwood Springs along the Colorado River, the coming of the railroad in 1887 encouraged commercial development, including new hotels, saloons, and other businesses. In the midst of this boom in growth, M. S. Yarwood opened Alexander's Cave in Glenwood Canyon. A diversion for the visitors who came for the hot springs, the cave opened early and closed late each day for leisurely touring. Unfortunately, the distance from the city discouraged many visitors from making the journey—especially with a variety of enticements close to the hotels, including the Yampah Spa Vapor Caves, which opened in 1893. Yarwood sold his cave to Frank Mason in 1893. Seeking more patrons, Mason renamed it Cave of the Clouds. He even gave customers free transportation to and from the cave on burros to boost his revenue. But his efforts failed; the cave closed by 1896.

West of Cave of the Clouds, The Cave of the Fairies above Glenwood Springs was opened in 1896 by attorney Charles W. Darrow. This cave benefited from stronger financial support than did Mason's Cave of the Clouds, and Darrow saw that a carriage road was built to its entrance overlooking Glenwood Springs. Armed with a larger marketing budget, The Cave of the Fairies boldly proclaimed through print advertisements its status as the eighth wonder of the world.

Darrow's employees made considerable efforts to open The Cave of the Fairies to the public. Enlarging undersized passages and filling yawning crevices with dirt and rock, the crew prepared a nearly level tour route to the cave's most attractive chambers. In 1897, at the tour's deepest point, miners excavated a tunnel that led to a fabulous view of Glenwood Canyon. This view soon became a tour highlight.

Manitou Cave just north of Manitou Springs was the last historic attempt at establishing a commercially operated cave in Colorado. Despite its convenience to the summer resort, it failed and closed less than three years after its July 1910 inauguration. Its demise was closely followed by that of The Cave of the Fairies in response to declining visits.

profit. The Boynton brothers and Charles Cross, who leased Cave of the Winds from owner Frank Hemenway, discovered their $1 admittance fee was too high. Facing an uncertain future, the operators walked away from their business in the fall of 1880, seeking employment elsewhere.

It took a major discovery by George W. Snider in January 1881 to revitalize the cave as a business opportunity. After digging his way into Canopy Hall and an extensive series of unknown rooms and corridors, Snider purchased the cave outright from Hemenway with partner Charles Rinehart and reopened it with an expanded tour in March 1881. Unlike the earlier tour that left visitors underwhelmed, the addition of Canopy Hall, Boston Avenue, and the Bridal Chamber convinced visitors a $1 admittance fee was a bargain.

OPPOSITE: Bill Allen illuminates eerie, damaged flowstone cascades in Cave of the Clouds, a commercial cave in the 1890s.

George W. Snider: Colorado's Cave Showman

George Washington Snider raised his six-shooter gun and pointed it at Frank Hemenway and his companions, ordering the men out of his cave. Snider was prepared to shoot and kill if necessary.

The men became suddenly silent after arguing with Snider over the possession of Cave of the Winds and backed off. Quietly, the group filed out of the cave, following the wooden stairs down to the floor of Williams Canyon. Snider propped the door shut to keep out the winter chill. He had survived the first big challenge to his ownership of Cave of the Winds. Resolving to pay Hemenway for the remainder of the purchase price, Snider set off for Manitou Springs.

It was January 1882. The previous January, he had pushed beyond a tight, dirt-filled passage in the cave to discover the wondrous Canopy Hall, an upper-level chamber that led to hundreds of feet of well-decorated passage. With Ohio lawyer Charles Rinehart as a partner, the 30-year-old stonecutter purchased the abandoned commercial cave and opened it for tours in March 1881.

Snider found his partnership with Rinehart uneasy from the beginning. They argued over money and the amount of work each put into the cave. Snider later recalled that he and his brother Horace performed most of the manual labor while Rinehart did little more than "lawyering."

Fortunately, Snider had an escape plan from his unpleasant business arrangements. In June 1881, he spent several hours excavating a crevice he discovered on the far side of Temple Mountain. A large plume of steam blew from it when he found it in December 1880, suggesting a hot spring within. But instead of finding hot water, he nearly fell into a large underground chamber when the floor of his dig abruptly collapsed. Gathering his candles, he carefully climbed down into the unknown cave.

The gloomy room was sizable, with corridors leading in three directions. Using lit candles as junction markers, he set off to explore. The central corridor led the farthest, to a huge, echoing chamber with beautiful stalactites and draperies. Realizing the cave might make his fortune, he retreated to the surface and carefully covered over the entrance. He would need to buy the property before any further exploration.

During the next three years, Snider spent his spare time working in the cave, preparing for the March 1885 opening of Manitou Grand Caverns. With considerable publicity, he opened the cave to enthusiastic visitors as the second underground destination in Manitou Springs.

On the other side of the mountain, George W. Snider's brothers and their business partner, Charles Rinehart, were unhappy. The new cave was certain to impact visits to their Cave of the Winds. Though they knew George Snider had been working on a new project, they were not prepared for the public interest surrounding its opening. Increasing its profile, Cave of the Winds opened Crystal Palace, a stunning new chamber featuring rare beaded helictites. Nevertheless, the number of tourists eagerly traveling the winding Cavern Gulch road to Grand Caverns and its view of Pikes Peak continued to grow.

Recognizing the public's affinity for the unusual, Snider featured in Grand Caverns a natural organ consisting of long draperies and stalactites that, when tapped delicately with a small mallet, played musical notes. Patrons greatly enjoyed the musical concerts in the Grand Concert Hall, often singing along to popular tunes. On occasion, musicians would bring along their instruments and accompany the natural organ to the delight of the visitors.

In July 1885, following the death of President Ulysses S. Grant, visitors began building in the cave's entrance hall the nation's first monument to the leader. General William Tecumseh Sherman visited the cave in 1889 and laid his own stone on the monument to remember his colleague. In honor of Sherman, accompanying visitors began building a monument for him. However, he kicked it over, telling them, "I'm not dead yet." Within a few years, Grant's monument grew to more than 10 feet tall and additional monuments were constructed for President Abraham Lincoln and the Confederacy's General Robert E. Lee. These three monuments still stand today.

A guide addresses a tour group in Cave of the Winds' Canopy Hall, discovered by cave developer George W. Snider in 1881.

Happy to get a little muddy, Joseph Blackshaw squeezes through a cluster of formations in Manitou Grand Caverns.

TThe photographer's assistants pose near a stairway in Manitou Grand Caverns' Canopy Avenue, 1890.
PHOTO BY WILLIAM HENRY JACKSON, COURTESY COLORADO HISTORICAL SOCIETY.

Perhaps owing to Grand Caverns' popularity, Charles Rinehart's wife, Rose, decided to bring George W. Snider to court in 1886. She claimed Grand Caverns crossed onto Cave of the Winds property. According to the business agreement between Snider and his former partners, all partners would jointly own any cave found on Cave of the Winds property.

During court proceedings in Colorado Springs the following year, an independent survey established that Grand Caverns did not encroach upon Cave of the Winds property. The jury decided in favor of Snider and dismissed the claim. However, Rose Rinehart's lawyer appealed to District Court. There, other surveyors showed the cave did indeed pass onto Cave of the Winds property, establishing that the older business owned half of Grand Caverns.

Because of an error by counsel, Snider did not appeal the case within 30 days, as required by law. Snider's lawyers argued the case should be considered anyway, as they indicated verbally within the deadline period that they would appeal. This appeal was denied.

In November 1890, Snider learned from several sources that the section corner and section boundaries had been tampered with prior to the second trial. Armed with evidence of criminal action, Snider appealed his case to the Colorado Supreme Court.

Although the first appeal was denied, the Supreme Court reconsidered and agreed to an 1893 hearing. Several witnesses reported to the court that

boundaries had been tampered with in violation of the law. After much deliberation, in a split decision the court decided to allow the District Court decision to stand—but only because Snider had taken so long to appeal. Since the District Court proceedings in 1888, both Charles and Rose Rinehart, two of the key witnesses, had died. The Supreme Court ruled it would be unfair to retry the case when the Rineharts were not around to defend their actions.

Lawsuits persisted through the 1890s as Snider continued to fight for the right to operate Grand Caverns. He attempted unsuccessfully to appeal the case to the federal courts. By the turn of the century, Snider had moved to California, as the joint operators of the caves believed it was best if he distanced himself from the property.

In 1915, Snider published a book titled *How I Found and How I Lost the Cave of the Winds and the Manitou Grand Caverns* in an effort to again become involved with the cave. Sadly, his brother Perry refused to sell the book in the cave's gift shop.

On June 26, 1921, 41 years to the day after John and George Pickett first squeezed into lower Cave of the Winds to rediscover the underground marvel, George W. Snider died in Los Angeles. His body was returned to Colorado Springs, where he was buried in Evergreen Cemetery. His relatives did not provide a memorial stone.

Redefining Cave of the Winds

After the 1907 installation of electric lights at Cave of the Winds, the cave's management decided to pave over much of the previously unimproved tour route, eliminating dirt and mud wherever possible. In portions of the narrow and winding cave, paved walkways now cover the entire floor. Although patrons in the early 20th century found this "taming" of a cave desirable, modern visitors have different sensibilities.

Current Cave of the Winds general manager Grant Carey, a former cave tour guide and caretaker, recognized that these early management decisions still affected his business. For the 1988 opening of the Adventure Room, Carey deliberately left the commercial trail dirt-floored in an effort to give visitors a sense of an undeveloped cave without concrete trails and steps. Although subsequent rainy years and deep mud have at times rendered the room impassable, Carey has taken an important step to present the cave as it was to today's audience.

Rather than limiting his horizons to the Cave of the Winds proper, Carey sought out other caves to open to the visiting public. In the early 1980s, an undeveloped caving tour allowed paying visitors access to Manitou Grand Caverns for the first time since 1906. This tour offered amateur cavers the opportunity to see not only sights that delighted Victorian-age visitors, but several chambers not previously shown. Heavenly Hall, discovered in 1989, proved the most popular of these new rooms. This chamber features an impressive display of beaded helictites.

By the summer of 1995, Carey converted the Grand Caverns tour into a lantern tour where underground travelers carry lanterns reminiscent of the glass ones used in the 19th century. Entering and exiting the cave through the excavated connecting passage from the Adventure Room in Cave of the Winds, tour groups are greatly entertained. Guides in period dress play the role of 1890s tour guides, bringing guests back to an earlier, more thrilling era.

The reopening of Grand Caverns to scheduled tours allowed Carey to redirect his caving tour to lower Williams Canyon's long-forgotten Manitou Cave. Though the cave had been closed since 1913, a few remnants of the previous commercial route remain, including an old pipe railing at the Whirlpool Dome and cleared walkways. Recent discoveries allow tours into a new section called Deepwater Cave, featuring long, low crawlways; towering domes; and a small, seasonal

Hazel Barton picks her way through the formations that cover almost every square inch of King's Row's ceiling, walls, and floor in Glenwood Caverns.

OPPOSITE: Paul Burger contemplates a stalactite display near Stone River in Breezeway Cave.

stream. Tours also frequent the Centipede section of Manitou Cave, crossing beneath the floor of Williams Canyon. When not closed by water, the connecting passage allows curiosity seekers to examine the cave's most historic section, discovered by local boys in the late 19th century.

🦷

Three months after watching cavers clear debris from the new platform in Glenwood Caverns' The Barn, I return in May 1999 to see how the commercialization process has progressed. With the cave's grand opening only a week away, I am interested in learning what has been accomplished.

By e-mail communication, manager Phil Kriz explained that the team still had much to complete. Yet, as I stand on the platform I earlier watched cavers construct by jackhammer, shovel, and hoe, I am amazed. A series of stairs and platforms defining the new route to King's Row stretches into the lower reaches of The Barn. Expertly placed lights illuminate much of the chamber, though at its bottom temporary work lights brighten the way for cavers jackhammering a constriction. Here, the stairs abruptly end, awaiting arrival of the carpentry crew.

Eager to see the construction up close, Ken Kreager and I descend the completed flights of stairs, stepping aside as other cavers carry up leftover scraps of lumber for disposal. The stairs and platforms follow the original caver trail, a steep, ill-defined pathway better suited to surefooted mountain goats than to humans. Undoubtedly, the new stairs will provide easier and safer access to the lower chambers, but at a cost some cavers believe is too high.

Like the taming of the surface wilderness above, opening a cave to commercial tours evokes conflicting emotions for many cavers. Although most appreciate the opportunity to share the beauty of the underground with a larger group, a strong feeling that caves can only be appreciated on their own terms persists. If that means climbing or squeezing or rappelling or even diving through flooded passageways to reach pristine chambers, so be it. Clearing trails and installing railings and electric lights all help make a cave more accessible, but these modifications also take away some of the cave's intrigue.

Reaching the bottom of the stairs and watching Colorado Springs caver Mike Frazier jackhammer out a large boulder,

I wonder if Steve Beckley's decision to open these chambers to the public was the right choice. For eons, this cave existed without human influence. Yet, in mere moments, Mike knocks out and shatters a boulder that had rested at the entrance to King's Row for a million years or more.

Collecting the pieces of the boulder and passing them up in five-gallon buckets to dump underneath the stairs, Colorado Springs caver Dan Sullivan helps Mike break out another rock. Within a week, hundreds of visitors a day will pass beyond this constriction on a new stairway that the carpenter is still planning. But none will know how this cave appeared when it first was discovered.

Volunteer cavers assist in the construction of security doors at Glenwood Caverns, 1999.
PHOTO BY STEVE AND JEANNE BECKLEY, COURTESY GLENWOOD CAVERNS COLLECTION.

I wonder how upper Glenwood Caverns, along the historic Cave of the Fairies commercial trail, looked to the original explorers a century ago. Passing a bucket back down to Dan for refilling, I also ponder how the original passageways of Cave of the Winds, where cement now stretches from wall to wall, appeared to George W. Snider and his brothers before the first visitors arrived.

Commercialization removes the wildness from a cave. But it also affords adventurers and tourists an opportunity to sample the beauty and wonder that many cavers take for granted. Are we cavers perhaps *too* protective of these underground worlds?

Pausing for a moment as Mike and Dan manhandle a particularly large rock from out of the dirt embankment, I remember that this opening to King's Row once required squeezing through a rectangular hole not much larger than my coffee table at home. When installed, the new stairway will allow even the most claustrophobic of souls to descend into the cavern and return with an ease unimagined by the cave's first explorers.

Perhaps, I decide, enjoying and appreciating the wonders of the room beyond are experiences best shared by many, not just a select few.

The Big Show

Despite a reluctance to generate publicity over caves, Colorado cavers and caves have been featured in news reports, television documentaries, newspaper and magazine articles, and radio reports. In 1986, the Denver Museum of Nature & Science (then the Denver Museum of Natural History) and Denver's Rocky Mountain PBS released *Silent Splendor*, a 30-minute television documentary on the 1984 discovery and subsequent study of the famous Cave of the Winds passage of the same name. Discussing the remarkable and rare beaded helictites, the documentary was uploaded to satellite and picked up by public television stations nationwide.

In 1988, the Denver Museum of Nature & Science and Rocky Mountain PBS teamed together again for *Lechuguilla Cave: The Hidden Giant*, a documentary on the discovery and exploration of New Mexico's spectacular cave near Carlsbad Caverns. Featuring several Colorado cavers who helped explore Lechuguilla, this documentary paved the way for the 1992 National Geographic television special *Mysteries Underground*, which also spotlighted the massive cave.

Nothing compares, however, to the March 2001 release of the large-format film *Journey Into Amazing Caves* by California's MacGillivray-Freeman Films. From the producer of the popular *Everest* and *Dolphins* large-format movies, *Journey Into Amazing Caves* is the most expensive caving movie ever produced. Budgeted at $3.9 million with funding from the National Science Foundation, *Journey Into Amazing Caves* brings the public into the mysterious world of caves.

One of the stars of the feature is caver Dr. Hazel Barton of the University of Colorado, a transplanted British microbiologist. A past chairperson of Denver's Colorado Grotto, Hazel has served as chief cartographer and an active participant in the surveys of several Colorado caves, including Glenwood Caverns, Fixin'-to-Die, Hubbard's, Porcupine, and the Clear Creek Fault Cave. Following a national search for cavers for roles in the feature, the film's producers selected Hazel and former National Speleological Society board member Nancy Holler Aulenbach of Norcross, Georgia, in July 1998. That September, the two cavers began a series of worldwide trips to exotic and spectacular caves as part of their filming duties.

Unaccustomed to the film industry's "hurry up and wait" nature, Hazel discovered filming can be a hectic experience. Fortunately, breaks in the filming schedule allowed her to begin a post-doctorate position in Boulder with noted microbiologist Dr. Norman Pace, another former Colorado Grotto chairperson. She also was able to continue her volunteer coordination of the ongoing Glenwood Caverns survey.

Microbiologist Dr. Hazel Barton, a renowned cave cartographer, surveys in Glenwood Caverns. She stars in the large-format film *Journey Into Amazing Caves*.

Unfortunately, Hazel's participation in *Journey Into Amazing Caves* brought criticism from some fellow cavers in Colorado and nationwide. Fearing damage and undue attention to fragile cave environments in the United States and elsewhere, these cavers inundated her and other film supporters with e-mail messages and telephone calls. Collectively, they criticized her decision to play an active and highly visible role in the feature. The National Speleological Society, however, has cooperated in the motion picture's filming, including assembling a scientific advisory board of noted cave scientists to ensure accuracy.

For their part, the producers pledged to exercise environmental sensitivity and to respect the caves in which they filmed. While filming underground, the company utilized improved equipment and techniques developed for its work on Mount Everest.

Without question, the large-format movie will continue to focus attention on caves and caving in the United States and abroad. Yet, with Hazel's participation in *Journey Into Amazing Caves*, Colorado cavers can rest easier knowing she played a role.

OPPOSITE: Bill Allen views the translucent calcite "angel wing" draperies, or "cave bacon," growing where water runs in a line along the ceiling of a Glenwood Caverns passage.

Cave
Conservation

At the beginning of the 20th century, the commercial Cave of the Winds near Manitou Springs had welcomed visitors into its meandering passageways for two decades. Thousands of men, women, and children had passed through the cave's entrance portal, admiring the delicate helictites of Crystal Palace and Dante's Inferno along with the large draperies of Canopy and Curtain halls.

Despite the cave's success, the condition of its interior concerned manager Charles H. Austin. Though he had served as manager for only a decade, he could see the damage that dusty trails and billowing smoke from the guides' kerosene lanterns and magnesium flares had wrought upon the cave's natural features. Beautiful white formations wore a uniform gray coat of dust, while hair and clothing lint hung from the walls like grotesque lacework.

With plans under way to install electric lights in the cave by July 1907, Austin asked employee W. O. Hooper to determine a solution for their dust and lint problems. Austin did not want the brilliant electric lights to disclose the cave's filth.

Hooper discovered that hand spray pumps filled with a dilute solution of hydrochloric acid could effortlessly wash down even the most delicate formations without harm. The effect was dazzling: Austin admitted to the *Manitou Journal* that he was pleasantly surprised by the success of Hooper's cleaning project. By April 1905, the entire tour route had

been cleaned and restored to the beauty that first engrossed the 1881 discoverers.

Curiously, Austin's delight at seeing the cave clean and lint-free did not launch a continuing maintenance program for Cave of the Winds. Though many commercial caves today have a regular schedule for cleaning passages of visitor-generated debris, Cave of the Winds proprietors waited until the mid-1980s to again clean their Discovery Tour route.

In the eight decades since Hooper sprayed the cave with hydrochloric acid, dust and lint had accumulated to significant amounts along the trail. Cavers working in Cave of the Winds on digging and surveying projects noted the unsightly buildup, recommending to management a major effort to restore the cave to its original beauty. In the fall of 1985, manager John Norton and general manager Grant Carey attended a cave management seminar at Carlsbad Caverns National Park, where they saw firsthand the remarkable restoration work in that popular commercial cave.

On a volunteer basis, cavers began clearing out piles of lint, spraying the walls and dripstone formations with water, and scrubbing them with toothbrushes. A large group spent several days in Curtain and Canopy halls washing the walls, bringing out colors not seen since the cave received its electric lighting system.

ABOVE: **A flowstone cascade in Cave of the Winds' Canopy Hall intrigues Chris Brown.**
OPPOSITE: **A limestone cliff holds the entrance to Cliffhanger Cave near Cave of the Winds.**

Noting the considerable time and effort needed to clean a single chamber by hand, cavers Walt Rubeck and Steve Sims suggested using a mechanical device for the remainder of the cave—a Hotsy® high-pressure washing unit. Used at industrial sites, the electric units heat water and add detergent to cleanse particularly dirty surfaces. In Cave of the Winds, however, only cold water was used to avoid damaging the cave environment.

Employing various levels of pressure from a narrow stream to a mist, the Hotsy® operators found they could wash everything along the tour route. Like a refreshing shower, the water scoured away dirt, lint, and grime from a century of visits—even revealing helictites previously buried in lint. With wet vacuums, the crew collected and dumped waste-water outside. By May 1986, John Norton's restoration project was completed. To the team's surprise, the entire route was cleaned in only six months.

As the cleaning crew progressed through Cave of the Winds, cavers provided assistance in other restoration projects. In the Oriental Garden, Walt Rubeck and Paul Burger painstakingly pieced together a long stalactite an electrician had shattered when installing new lights in the late 1950s. Using a variety of glues and cementing compounds, Walt found a combination that would hold the massive formation together. Unfortunately, within a few months, two off-duty guides accidentally bumped the repaired formation, knocking it to the floor and shattering it again. This time, the cave handyman drilled a hole through the stalactite and installed a piece of rebar to avoid any future breaks.

Repairing Damage

Throughout the United States, cave restoration is a growing trend, both in commercial and undeveloped caves. Littleton caver Al Collier has coordinated recent efforts to repair stalactites, stalagmites, and other cave formations in private and public caves throughout Colorado.

Piecing together fragments of broken stalactites and stalagmites like a jigsaw puzzle, Al has successfully repaired and restored to near-original condition many dripstone formations in the caves of Williams Canyon. Using carefully selected glues, Al repaired badly broken formations outside the cave, then returned them for final repair on-site. During a two-year period beginning in 1998, Al repaired formations in Cave of the Winds, Breezeway Cave, and Huccacove Cave.

OPPOSITE: Soda Straws and stalactites dangle from the ceiling in Cave of the Winds' Oriental Garden.

A family marvels at the magnificent Oriental Garden and Mirror Lake in Cave of the Winds, the site of two major cave-cleaning efforts.

Other Colorado caves have seen Al's careful efforts to repair and restore damaged formations and passageways. In seldom-visited Buffalo Cave near Dotsero, Al discovered a severely damaged wall of moonmilk, a soft, plasticlike deposit of carbonate minerals. Spending several hours at the task, Al managed to erase the writing from the wall, though not completely repair the moonmilk.

In many caves, graffiti is a continuing problem. Many once-pristine caves now bear scars from the selfish need of some individuals to leave their names for future visitors to see. One of the state's most graffiti-ridden caves is Huccacove Cave, the now-closed commercial cave in Williams Canyon. Penciled, smoked, and painted graffiti covers many of the walls and ceilings in the cave's historic section, open to all comers for a century. Beginning with the gating of the cave in 1978, cavers from the Southern Colorado Mountain Grotto and Colorado Grotto have spent hundreds of hours removing and hiding graffiti. In some cases, the graffiti clings to a layer of dust that can be wiped or scrubbed clean. Colorado Springs caver Marc Hament discovered that mud can be used almost as a plaster of Paris to cover the graffiti. In nearby Manitou Cave, Marc used clay he had collected to hide unsightly writing from the view of participants of the commercial Explorer's Tour. Marc also employed this technique in 1997 to hide spray-painted writing and arrows in Fly Cave near Cañon City.

Sometimes the carelessness or laziness of one group can help preserve cave features for future recovery. Teams of volunteers have carefully removed elevator debris dumped in the old lunchroom in New Mexico's Carlsbad Caverns, revealing delicate rimstone dams and other speleothems. Golden caver Linda Dotter spent most of a day in Colorado's Glenwood Caverns digging out refuse dumped in a small alcove by 1890s miners excavating a tunnel to the spectacular Exclamation Point. As she reached the original rock floor, she found fragile stalagmites that visitors would have probably damaged

Wielding a spray bottle and toothbrush (top), Al Collier painstakingly scrubs formations touched by visitors to Glenwood Caverns and (above) pieces together one of hundreds of formations he has restored during his conservation work in Colorado caves.

or removed during the 40 years the cave stood unguarded. A decade earlier, in Cave of the Winds, cavers spent a day removing dirt and rock dumped at the base of the large Atlas Column, uncovering flowstone and a small rimstone dam buried during the Canopy Hall trail construction.

Current Practices

Conservation-minded caving has gained speed during the last couple of decades. Since the mid-1980s, cavers increasingly

lay taped trails through delicate passageways, restricting future travel to a single route. Taped trails lead through many of the larger rooms in Williams Canyon's Breezeway Cave. In Cave of the Winds' Silent Splendor, blue reflective markers and aluminum posts increase the visibility of the taped trail and discourage photographers from leaning over for a closer view of the passage's remarkable beaded helictites.

Exploration or survey teams often leave tapes with written messages in sensitive passages. The "boots off" message commands arriving cavers to remove muddy footwear before traversing the upcoming passage. "No going leads" encourages cavers to avoid crossing the tape and, instead, to more closely examine apparent leads.

Larger, more complex caves may even feature color-schemed tapes, with certain colors marking leads, main routes, and other passageways. In the mid-1980s, Forest Service volunteers installed Popsicle sticks with different colored reflective tape in Groaning Cave to mark the main route into and out of the cave. Although the plan aimed to keep cavers from venturing into less-visited parallel passageways off the major trade route, many cavers cried out against the markers, declaring the reflective tapes as damaging to the cave's "wilderness experience." Within a year, all markers had been removed.

Cavers also introduce artificial aids to reduce impact on a cave. In Breezeway Cave, a bridge replaced the delicate boots-and-muddy-clothes-off traverse on a narrow ledge above the Holy Waters pool. This new bridge protects the pool from any possible falls of passing cavers or from mud lumps dropped into the water. Golden caver Tom Dotter installed an environmental gate in the nearby Antelope Freeway, an excavated crawlway leading to many of Breezeway Cave's larger chambers.

OPPOSITE: Steve Lester chimneys up a canyon in Breezeway Cave as Bill Allen watches out for the climber's safety.

All the Right Moves

Experienced cavers understand the techniques necessary to move through caves with care. Unlike a bull in a china shop, cavers can pick and choose a route, minimizing their impact on a fragile environment.

In highly decorated cave passageways, conservation-minded cavers sometimes take several minutes to negotiate only a few feet. Moving an arm or a leg at a time, a caver can appear nearly motionless while the next move is determined. Sometimes cavers select an alternative route to avoid a particularly pristine or delicate area.

Cavers usually watch their companions as they pass through decorated regions. They help direct each other's movements with verbal commands, treating the cave as a prized personal possession.

Speed is an enemy of cave passageways, for fast travel encourages careless motions. Ask any caver who has chanced upon a cave and then seen it degraded by even experienced visitors within five or 10 years of its discovery. Heedless, hurried travelers can damage or destroy passages beyond recognition.

Responsible cavers take particular care around bodies of water, as these often harbor microscopic life. In caves where water pools rather than flows, thoughtful cavers don't draw water, wash hands, or otherwise disturb these underground reservoirs. In some caves, sterile containers are used to transfer water from pristine pools to caver water bottles to avoid contaminating the water supply.

Carl Bern cautiously ascends a flowstone-covered passage in Glenwood Caverns' Beginner's Luck, careful not to disturb the plentiful formations.

When entering previously unknown passages and chambers, lead cavers exercise utmost caution in determining where to walk. Even pristine mud or dirt floors have value, not only to future scientists who study the room, but also to visitors seeking a wilderness experience. Often, each team member follows a single narrow trail or even the first caver's footprints to avoid impacting the cave more than necessary.

Cavers use discretion when discussing caves and cave features. Because some uninformed or uncaring members of the public view caves as nothing more than holes in the ground, it is often better to say nothing of them to those outside the caving community. Many experienced cavers believe caves should never be discussed or publicized for any reason. Sadly, the continuing destruction of once-wonderful caves gives credibility to this opinion.

Concerned conservationists can help combat the desecration of our nation's underground wilderness by actively supporting caving and conservation organizations, removing trash from caves, and notifying cave owners or administrators of vandalism. On a larger scale, karst regions with sinkholes and cave features also need safekeeping. Cavers have discovered trash and pollution in caves many miles from the entry of a stream. In some cases, caves can become too contaminated to safely visit.

Much like our nation's surface wilderness, our underground lands are certainly worthy of protection from lasting human impact. Though cave conservationists' efforts to establish formal underground wilderness in federal lands have so far been unsuccessful, the concept is worthy of continued discussion and consideration.

Caves are nonrenewable resources, at least during the lifetime of humankind. As stewards for future generations, we must vigorously defend caves from any abuse or damage.

Bill Allen gazes into Holy Waters in Breezeway Cave.

The crawlway had altered the airflow through the cave, increasing wind speed and drying out adjacent rooms. Tom's son Matthew studied the problem and then offered the environmental gate solution as part of his Boy Scout activities.

Environmental gates have also been installed in Cave of the Winds to limit airflow. Majestic Hall along the Discovery Tour route became noticeably moister following the discovery of the Whale's Belly and Silent Splendor, two wet and muddy rooms that had been sealed off from the rest of the cave for millennia. After the installation of environmental gates at the base of the Whale's Belly and at Four Fingers Junction, Majestic Hall returned to its dry and dusty state.

Gating caves to control visitor access is a common practice in modern cave conservation. Constructed of steel alloys and of sophisticated designs, current cave gates see little of the damage that their 1970s predecessors experienced. Beginning with the 1978 gating of Huccacove Cave in Williams Canyon, cavers fought locals who believed the privately owned cave should be open to everyone. Several times, cavers found the gate damaged or completely removed.

Property Rights

Unfortunately, many would-be cave explorers do not recognize the rights of private or public property owners. Cave of the Winds has issued trespassing summonses to unauthorized visitors to their posted property. A former caretaker took property protection even further by pointing his rifle or tossing dead skunks and raccoons at trespassers.

Trespassing has continually plagued the caves of Williams Canyon. Following the closure of Manitou Grand Caverns in 1907, the Cave of the Winds staff found the tunnel entrance gate broken on numerous occasions. The management of Manitou Cave discovered one morning in 1912 that vandals had broken into the cave and stolen all the lightbulbs. During the next 40 years, the Cave of the Winds management sealed various Williams Canyon caves to keep out visitors only to see them reopened through unauthorized digging efforts.

In early 1885, George W. Snider, the developer of Cave of the Winds and Manitou Grand Caverns, worked with El Paso County State House Representative Charles W. Barker to submit to the Colorado Legislature an act protecting Colorado caves from vandalism. This effort was enacted into law on April 9, 1885, as the nation's first cave protection act. Although

By 1958, the Cave of the Winds management installed chicken wire and stern warning signs to discourage vandals. Unfortunately, the cave already suffered damage from the previous 60 years of visits. PHOTO COURTESY CAVE OF THE WINDS COLLECTION.

this historic law was successfully used in August 1949, when a local court fined an Iowa man $37.80 for breaking a stalactite in Cave of the Winds, it was dropped from the Revised Statutes on July 1, 1972.

The 1988 Federal Cave Resources Protection Act provides protection and additional management responsibilities for federal caves. Signed into law by President Ronald Reagan in January 1989, the act calls for federal agencies to designate caves as "significant" in order to receive increased protection. With limited funding, federal agencies have been slow to designate significant caves on their lands and provide additional management of the resources. Colorado's White River National Forest, in its Revised Land and Resource Management Plan, has designated several caves as significant, including the popular Fulford, Hubbard's, and Spring caves.

Many cavers believe that through the continued education of the general population about the delicate nature of caves, visitors underground will tread more lightly. Just as growing numbers of Americans recognize bats as helpful animals that consume great quantities of insects—not the fearful creatures depicted in Hollywood films—there is hope that the public will learn to see caves as also worthy of care and protection.

Caves are a finite resource. Like wilderness, few unspoiled caves remain to be explored and treasured. Decisions we make today will affect future generations and their underground experiences.

CHAPTER 7

Photographing in the Darkness

BY
David Harris

It took almost two hours of hard caving to reach the spot where Steve Beckley, Bill Allen, and I now stand. We shoved boxes of flashbulbs, packs loaded with camera equipment, and our bodies through tight passages to reach King's Row in Glenwood Caverns. As the developer of this wildly beautiful cave, Steve asked me to make a series of photographs to advertise his longtime dream. In a matter of weeks, an entrance tunnel will be completed here, allowing anyone to walk inside the cave and see for themselves what I hope to record on film.

As Steve heads back to the surface, Bill, a fellow caver who often assists me with my photographic endeavors, helps set up the tripod and camera equipment on a wet, glistening, flowstone-covered slope. It takes only a few minutes to arrange the complex lighting system of electronic flashes and old-fashioned flashbulbs. As Bill positions himself for human scale in the photograph, I shoot the scene on a sheet of Polaroid film to assess the exposure and composition. Starting the timer, I pull the test photograph from the camera and wait for it to complete its two-minute development cycle.

Instead of talking with Bill as usual while I wait, my mind wanders back to the time and place where my fascination with cave photography began. I fondly recall my tiny Kodak 110 camera I took on vacation with my family 22 years before.

I was a high-school student then and my family was visiting Cave of the Winds on a summer trip. With a pocketful of magic cube flashes and two packages of film, I was the child with his hand in the cookie jar. As we walked through the cave, I snapped photos of the pretties and aimed my camera into the passages that led off to dark halls, hoping my flash would reveal what I couldn't see.

Later that day as we drove home, my mind buzzed with what I had experienced in those caverns. Something had reached in and sparked a curiosity that begged to be satisfied. Perhaps my tiny, undeveloped photographs could answer that need.

The good and bad of getting film back from the processor is that either great surprises or real disappointments hide inside that bright yellow folder. I experienced both. The photos of the formations, the close subjects, pleased me. I was deflated, though, to find I could see nothing more in those dark, beckoning passages. Even though the pretty formations graced the pages of my photo album, they only confirmed what I already knew: I captured their images in a way that recalled just how I saw them. I had documented and stored them as "memories."

But the dim, disappointing images had captured my imagination. I had visualized much more than I was able to demonstrate with the camera, and—as Cave of the Winds was now 2,000 miles away—there was no going back to try again. The thrill of the unknown, the power of so many questions

ABOVE: Hazel Barton negotiates Lower King's Row, an area visible from the visitor trail in Glenwood Caverns.
OPPOSITE: Sunset touches the hills beyond the entrance to Indian Cave, a White River Plateau shelter cave.

The famous Cathedral Spires in Cave of the Winds' Bridal Chamber were the cave's advertising symbol, circa 1885. PHOTO BY WILLIAM HENRY JACKSON, COURTESY COLORADO HISTORICAL SOCIETY.

caving: climbing, digging, rappelling, exercising the principles of conservation, and using the proper equipment. But his instruction stopped at showing me how to photograph what I saw. That was all up to me—and that was exactly how I wanted it.

Membership in the NSS provided access to many publications and contacts in the caving world. Each month, a thin journal called the *NSS News* arrived in the mail. I studied the photographs of others, noted who they were and what equipment they used, and absorbed everything possible about caving and underground photography and never ran out of room to learn more. My curiosity was insatiable but my skill in making the kind of cave photographs I wanted was abominable. Disappointment was out of the question: There'd be no slowing down until I made the kind of cave photographs I imagined, and tried to make them better than any I had seen.

Joe Ondrachek used a large-format camera and a tripod to photograph Cave of the Chimes, circa 1955. PHOTO BY JOHN STREICH, COURTESY JOHN STREICH COLORADO GROTTO COLLECTION.

Armed with a good flash and a relatively inexpensive 35mm camera, I set about making photographs on every grotto trip I could attend. My pack, which once carried only caving essentials, grew to about twice the size of everyone else's. I had already determined that simple snapshots would never do: I pushed the limits of my equipment with elaborate ideas and tested the patience of my fellow cavers, who just wanted to go caving. I needed to strike a balance between the demands of photography and the desires of the cavers I depended on for assistance.

Beginning to recognize what makes a good cave photograph and what doesn't, I often played a game with myself as I caved with my friends. At every pause, every food break, every wait while someone checked out which way to go, I mentally set up a photograph. I found the right angle for my subject, placed a caver for scale, and positioned the flashes to

rolled up into an intangible feeling, manifested itself in those photographs. I started then and there to learn everything I could about caves. Seven years later and finished with college, that feeling would steer me back to Colorado.

Unraveling a Mystery

Colorado has more than 250 caves, according to Lloyd Parris' 1973 book, *Caves of Colorado*. In the summer of 1984, I carried my freshly purchased copy from the Chinook Bookshop in Colorado Springs and devoured the text and photographs. I also noticed references to the National Speleological Society (NSS) and caving clubs in Colorado. Soon I befriended the contact person for the Southern Colorado Mountain Grotto in Colorado Springs, Todd Warren.

I jumped right into caving with Todd. On some amazing trips during the next two years, he taught me the essentials of create the effect I envisioned. The photographs that came back after each trip seemed to demonstrate my increasing aptitude for finding good locations. I wasn't, however, getting something right. No matter how incredible the location was, if some key ingredient wasn't present, the photograph didn't convey the right feeling.

Chomping at the bit to create those smashing images that a few seasoned photographers were routinely making, I realized I still didn't have the skill. What did they know that I had not yet learned? Why was this so difficult? My answer came from other cave photographers. It all had to do with light.

I set about cracking the secrets of lighting that other cave photographers knew and, luckily, they left plenty of clues. Critiquing every photograph I could find, no doubt boring my family with my incessant analysis, I had finally arrived at a conclusion: The images that worked best were all lit with a "natural" light, meaning sunlight. Although many sources could light a scene, the images that really stood out had a quality that mimicked one main light source, just like the sun. This epiphany completely redirected my photography, which now required much greater commitment.

I put down the 35mm camera I had been using, as it was time to learn about light.

Counting my savings and coming up short, I sold my truck (and a few other things I probably shouldn't have sold) and bought a medium-format camera. Entirely mechanical, it had no batteries to change and no electronics to go haywire. It was big and heavy—and it was crazy to take such a camera underground. Although my pack quadrupled in size and weight, my health and cave training allowed me to still keep up with my friends. That camera could do what most couldn't: I replaced the regular film holder with a Polaroid film holder for a big, 2¼-square-inch image that was easy to view. Now I could study lighting in the cave, make adjustments, make more adjustments, and crawl out confident that I had what I wanted. No more second trips to faraway places at great expense. This knowledge went hand in hand with carefully choosing caving companions, for the time involvement made it necessary to take a "photo trip" rather than a traditional caving trip. I refined my lighting technique and came to the conclusion that the best cave lighting ever invented is the flashbulb. It's a little piece of the sun for the shadows underground.

William Henry Jackson, one of Colorado's first cave photographers, captured this view of Williams Canyon from the Temple of Isis around 1882.
PHOTO COURTESY COLORADO HISTORICAL SOCIETY.

A modern-day look at Williams Canyon from the Temple of Isis.

One of the most-photographed rooms in Cave of the Winds was the Bridal Chamber, known for its Cathedral Spires, the stalagmites to the left. The early electric lighting system is visible in this view, circa 1920. PHOTO COURTESY CAVE OF THE WINDS COLLECTION.

Although the Broken Column, obviously pieced together, is now missing from Manitou Grand Caverns, the view of Horseshoe Tunnel looks much the same as it did in 1885. PHOTO BY WILLIAM HENRY JACKSON, COURTESY COLORADO HISTORICAL SOCIETY.

Old-time Lighting

Photographers have been experimenting with different ways to illuminate the darkness for more than a century. Beginning in 1882, photographers enjoyed some success in illuminating Colorado's few commercial caves. It's not hard to imagine how reluctant an early photographer would have been to carry an 8x10 view camera into the damp cold of a cave. It's even harder to believe anyone would want to spread a volatile liquid emulsion, called collodion, onto a piece of glass that would become an original negative only with careful and persistent coaxing. Charles Waldack did just that in 1866 at Kentucky's Mammoth Cave, using magnesium powder as his light source. His are the first high-quality photographs produced on "wet" plates to advertise a cave. The highly explosive flash powder he used would remain a primary light source for photography above and below ground until the early 1930s.

William E. Hook made one of the earliest known underground photographs in Cave of the Winds around 1885. Hook produced his "limelight" by using a burner that mixed bottled oxygen and hydrogen together in a flame directed against a lump of calcium carbonate. The reaction was very bright and almost smokeless, but photographic plates were quite insensitive to the color of light it produced. Long exposures were complicated by the danger of serious explosions. For all his effort, Hook's photographs were generally of poor quality.

Some consolation arrived with the availability of early versions of commercially prepared "dry" film, which replaced the tedious, hand-prepared, wet glass plates.

William Henry Jackson knew about the Williams Canyon caves and had made numerous photographs at Cave of the Winds and at the Manitou Grand Caverns back in 1882. As his lighting choice, Jackson probably favored magnesium; though it was more expensive than limelight, it was much easier to handle. Other examples of early photographic efforts in Hubbard's Cave and Fairy Cave (now Glenwood Caverns) at the turn of the century all were produced with similar light sources.

These early photographers probably packed more than 100 pounds of equipment into the caves and took most of the day to get one good shot. The flash of magnesium powder smoked up the cave so badly it was hard to find the way out. It would be 1929 before the invention of flashbulbs, which made the cave photographer's life a lot easier. Even though they occasionally exploded, flashbulbs were smoke-free and a great convenience when combined with the smaller cameras and improved film available in the 1930s. Flashbulbs are increasingly expensive because they're no longer produced today. Even so, their quality of light, small size, and ease of use make them the next best thing to the sun for cave photography.

Taking the Challenge

Changing the way I manipulated light in my photographs made all the difference. The quality of light the flashbulbs provided and the way I started positioning that light were the intangible elements I had been missing. Now I possessed the complete alphabet for the language I wanted to speak. I could write a thousand words with one carefully considered press of the shutter. People started to notice what I was saying.

Large, highly decorated rooms such as Temple of Silence in Cave of the Winds require multiple light sources to create a natural effect; an amazing eight flashes were used to make this photo.

Two Flashes Are Better Than One

Making photographs underground proves an exciting opportunity for anyone with a camera. The challenge lies in how much effort the photographer is able to put into composing and lighting each scene. For most, the reality of cave photography centers on a trip into a commercial cave; guided trips move quickly and access to anything off-trail is usually prohibited. But these restrictions shouldn't discourage shutterbugs from making some great cave photos along the way. Here are some simple steps to get you started:

1. Begin by using a basic on-camera flash, ideally with through-the-lens metering. This setup provides a great way to document a cave trip and gives you a feel for what your system can do.

The artificial lighting in the majority of commercial caves can be shot over— that is, virtually eliminated—with most electronic flash systems. The average cave tour allows little time to set up a tripod, and most lighting systems installed in caves cause horrible color problems, so your electronic flash offers a creative advantage. Besides, on most cameras, when the flash is turned on, the camera automatically switches to a shutter speed of 1/60th of a second or faster, which makes a tripod unnecessary.

It's not essential to buy 1600-speed film when good-quality, 200-speed film works great with an electronic flash. Use color print film, as it can "see" into darker areas better than slide film.

2. After photographing your first cave and reviewing the results, take a trip with a friend who can carry a second flash unit for you. This off-camera flash will need a "slave" to trigger it. (Sorry, your friend will never do as a slave!) Photo slaves are devices that attach to the off-camera flash with a short wire; the slave and second flash can be found at any decent photography store for about $20.

To photograph each scene, position the off-camera flash anywhere between five and 20 feet down the trail from the camera, and point the slave toward the on-camera flash. The off-camera flash should be aimed into the center of the same scene as the camera. Firing the camera and attached flash into the scene as usual should automatically trigger the off-camera flash at exactly the same time.

Because the off-camera flash won't receive the same light-balancing signals as the on-camera flash, manual adjustment of the off-camera flash's light output will be necessary. Here is where some experimentation really pays off. The beauty of this two-flash system is that both photographer and helper can move along with any cave tour, but still have an outstanding chance of capturing some fine photographs.

3. When it comes to cave photography, experience is the best teacher, but the willingness to constantly experiment is essential. As a final tip, try using an off-camera cord to get the attached camera flash away from the camera, too. If your lighting starts to resemble the quality of natural daylight underground, then you just might be on to something!

In Glenwood Caverns, Hazel Barton walks among the aragonite crystals of Paradise.

A telephone call came from New York one day in 1995. An editor with *Air & Space/Smithsonian* magazine wanted to see my cave photography. NASA was sending three scientists into New Mexico's famous Lechuguilla Cave as a precursor to a search for life on Mars. *Air & Space* wanted coverage of the scientists' cave trips, and the editor decided I was the one to document their explorations of a cave almost 100 miles in length.

Perhaps, I thought, as I prepared my equipment for the two-week assignment, it really was crazy to carry this gear underground. Spending five days in the cave at a time, coming out, going back underground for five more days—not quite what William Henry Jackson had done in Cave of the Winds. I supposed, however, that I had a good chance for success. I wasn't permitted to have an assistant to help carry my 80-pound pack of tightly compressed food, sleeping gear, caving essentials, and specialized photography equipment, but it didn't matter. I had long prepared myself for this challenge.

Days later, I climbed the last rope out of Lechuguilla Cave into the sunlight I had so missed. I carried everything out that I had carried in. Cave photographers have a saying that exposed film is much heavier than the unexposed film they start shooting with. So much time and effort go into making each image that some transference of energy as weight must exist. Surely Albert Einstein would have agreed with this saying. However, if this were the case, then I couldn't have carried my load back to the surface. I did, and my pack seemed to weigh only 20 pounds as I reveled in the thought that I had captured a set of historic photographs of the sunless days those scientists spent underground. After all, I had refined my technique during many other cave trips, used all the resources my 140-pound body could muster, and done my best to emulate the natural light I now stood under. The few Polaroid sheets I had room to carry told me I had at least gotten as close as ever to my photographic goals. The "real" film would be the proof.

The beeping of my wristwatch's countdown timer jars my thoughts, and suddenly I return to the absolute darkness of Glenwood Caverns. It's time to pull away the protective paper covering the Polaroid proof of the scene in which Bill Allen waits patiently. He hasn't said a word in the last two minutes and I silently thank him for my moments of solitude. What an odyssey it has been with this difficult passion called caving. What a blessing it has been to crawl out from under so many rocks to the light of day and follow my dream of understanding what lies in the darkness beyond my flash. I had learned from the disappointments and had followed my curiosity to this exact spot.

The Polaroid proof reminds me of my long journey: It came out better than I imagined.

OPPOSITE: Lighting is a crucial element in the success of an underground photograph such as this one of Steve Lester and a helictite colony in Breezeway Cave's Elkhorn Chambers.

Visiting Underground Worlds

Picture yourself deep beneath the surface of the earth. With only the electric light on your helmet to illuminate the alien landscape of tans, reds, and grays, you follow your companions through a tube, stooping to avoid bumping your head on the low ceiling.

Taking care of your footing on the slippery surface, you see the tube leads sharply down into a noisy chamber. A constant roar necessitates shouting to your cohorts. As you slide down, you discover the roar is from a small stream that flows through the corridor, dropping over small waterfalls and frothy cascades.

According to the photocopied Fulford Cave map that someone in your party produces from his pocket, the passage to your left ends in only 50 feet. To your right, however, several sizable rooms are indicated.

You decide to follow the stream to seek its source. Ahead, a boisterous band of Boy Scouts bounds from rock to rock, flashlights illuminating both the corridor and each other. "There's a big waterfall ahead," several scouts excitedly inform you.

You pass the group, gaining elevation as the corridor leads to another junction. Here, a steep, muddy ramp leads up and out of sight to your right. The stream emerges from the passage to your left. Curious about the waterfall, you follow the left branch.

Within only a few feet, the stream, never very deep, spreads out to cover the entire floor. Carefully, you pick your way from rock to rock, trying to keep your boots dry. Though you're muddy from your trip into the cave, you know wet boots can make for an uncomfortable trip out.

Around a bend is the waterfall the boys reported. It's not very big, perhaps six feet or so in height, but impressive nonetheless. One of your companions pulls out a camera from his coat pocket. "Let's have a group photo," he suggests.

As you crowd together in front of the waterfall, one of your boots slips into the water. Fortunately, the cold, clear water doesn't reach the top of it. You hope the waterproofing you applied last winter keeps the water out.

A bright light flashes from the camera. The photographer takes another to be sure.

"It's nearly 3 o'clock," someone announces, glancing at his wristwatch. "Maybe we should head out."

Because it took the group more than an hour to reach this point, you reluctantly agree. But, you resolve, you'll return to explore that right-hand climb and the cave's upper level you heard about from a group of departing cavers at the entrance.

To your surprise, the adventure and the mystery of a Colorado cave have been an intriguing experience, despite your initial fears and uncertainties. For some, a journey into an undeveloped cave like Fulford can fuel a new passion for the darkened lands beneath our feet. For others, an excursion

ABOVE: Snowmelt helps feed this Fulford Cave waterfall, one of the cave's main attractions.
OPPOSITE: Administered by the White River National Forest, Fulford Cave is Colorado's most popular undeveloped cave open to the public.

Underground visitors Eric Crandall and Tara Foote enjoy the historic Bridal Chamber on Cave of the Winds' Discovery Tour.

On the Cave of the Winds' Explorer's Tour, cavers are invited to try their hand at digging for new passages.

into a well-lighted commercial cave like Cave of the Winds is all the excitement necessary to satisfy their curiosity.

Colorado's Commercial Caves

The great majority of the visitors to Colorado's underground labyrinths tour the commercially operated Cave of the Winds and Glenwood Caverns. Every year, more than 230,000 people travel these caves' well-illuminated passageways in the company of trained guides who offer a moment of total darkness as a break from the steady stream of history lessons, geologic explanations, and safety directives.

At Cave of the Winds, the Discovery Tour leads groups through a confusing series of passageways that crisscross each other at varying levels. The tour visits many of the chambers George W. Snider found in 1881, in addition to rooms in the former Middle Cave, connected to Cave of the Winds by Snider's nephew Ben in 1929. Ben squeezed through a tight hole and a pool of water known as the Rat Hole; today this squeeze, renamed Fat Man's Misery, has been considerably enlarged.

The largest rooms on the tour are Canopy Hall, featuring a wishing well and time capsule; Temple of Silence, home to numerous stalagmites and stalactites; and Adventure Room, where tours turn back for the entrance.

On the western side of the state, tour groups visiting Glenwood Caverns first follow the historic Cave of the Fairies trail through the cave's upper level. Though vandalized during its 40 years of abandonment, the passage is still in remarkably good shape.

The best is yet to come, though. Following a stop at the Exclamation Point view of Glenwood Canyon, the tour reverses course and follows the same passageways back to the entrance. Here, a guide leads sightseers down a steep pathway to the lower-level entrance to The Barn and King's Row, the cave's two spectacular showpiece rooms. Known only since 1960, these chambers are still active, with moist, dripping stalactites and wet, colorful flowstone, expertly illuminated with carefully placed lights. The darkness crowds the lighted pathways, giving the visitor a sense of exploration.

For the more adventurous, both Glenwood Caverns and Cave of the Winds offer alternative tours. A longer caving tour at

OPPOSITE: In Glenwood Caverns, Bill Allen beholds one of Colorado's largest known draperies.

Glenwood Caverns provides an opportunity to explore beyond the lighted pathways. The tour follows meandering passages from the upper-level tour route, taking in chambers previously unseen during the cave's early commercial period. Participants must be in good physical shape, however, as the route is frequently low and crawly, with ample opportunity to get dirty.

At Cave of the Winds, The Explorer's Tour voyages into lower Williams Canyon's long-closed Manitou Cave. Requiring plenty of climbing and crawling, the unlighted route leads precipitously down to one of the former commercial tour's attractions, the Whirlpool Dome. From there, the visitor has a choice of routes. One leads under the canyon floor through a low, muddy crawlway to the historic Centipede Cave on the canyon's east side. Two other passages lead up and through a jumble of rocks to the Deepwater section of the cave, where cavers are still making discoveries. Tour groups are allowed to visit these new passageways and even try their hand at digging.

For those who prefer not to get grimy on their trip underground, Cave of the Winds offers a less strenuous trip into the historic Manitou Grand Caverns. Led by guides in late 19th-century dress, the trip includes handheld kerosene lanterns to light the many large chambers along the route. One of Colorado's biggest rooms to date, the Grand Concert Hall is a highlight of the journey. Depending on trail conditions, tours also drop down into the Fairy Bridal Chamber, where weddings were held by torchlight as early as 1892.

For more information on touring Colorado's commercial caves, please see Appendix A.

Popular Undeveloped Caves

Most of Colorado's caves are on public lands administered by the U.S. Forest Service and the Bureau of Land Management. Though some are closed and gated by administrative order, many are open for skilled visitors with the training to safely explore undeveloped caves.

Central Colorado's White River National Forest contains three favorite wild caves. Fulford, Spring, and Hubbard's caves are identified on Forest Service, USGS topographic, and some state road maps. Of these, Fulford attracts the largest number of amateur cavers, probably because of the gentle, Depression-era Civilian Conservation Corps trail to its maintained entrance and the Forest Service campground at its trailhead.

Of Fulford's three entrances overlooking the Brush Creek valley, the central culvert entrance sees the most traffic. This is the entrance that miners opened in 1893; Forest Service volunteers replaced the wooden timbering in 1986. Fulford comprises about a mile of passage on three levels, including several large, impressive rooms. However, the underground stream and waterfall are the main attractions for most visitors, who follow the passages up and over the Devil's Washboard to the stream gallery.

Some explorers climb through an obscure ceiling hole above the Devil's Washboard to the cave's middle level, where the Breakdown Room provides access to smaller, decorated chambers. Exercise caution in this cave, as several deep, dangerous pits connect the various levels; unlucky and careless visitors have fallen numerous times in recent years, requiring assistance and extrication by search-and-rescue teams. Inadequately prepared visitors have gotten lost in the cave, or run out of light. One such group in 1991 burned their cash to stay warm while awaiting rescue!

Along the South Fork of the White River, another Forest Service trail leads up to the large entrances of Spring Cave. Excursions to the cave are particularly enjoyable during the early autumn months, when the aspens of the South Fork blaze with brilliant yellow and red.

From its spacious entrance rooms, where the Colorado School of Mines Student Grotto coordinated its November 1962 expedition to explore and map the cave, descending passages lead down a wooden ladder and a questionable Manila rope to Thunder Road. This is one of the largest underground streams known in Colorado. In the late spring and early summer, it can fill all of the cave's lower-level passageways. Twice during recent years, the stream overflowed the lower levels and drained out of the cave's entrance. Therefore, use proper care in visiting the cave during the early season, the time of year when snowmelt runs highest on the White River Plateau.

During late summer and autumn, it is possible to drop down into Thunder Road and follow the stream for 300 feet. Near the junction with the entrance series of corridors, another passage heads around a corner to a beautiful emerald-green lake.

Knowledgeable cavers can find an obscure hole that drops them down deeper into the cave, passing through Butterscotch Passage to the Bridge Room and several other

OPPOSITE: Ice stalactites and columns surround Bill Allen and L. P. Lawrence in Fulford Cave.

VISITING UNDERGROUND WORLDS   117

Red-carpeted steps lead to the Cave of the Winds' Bridal Chamber.

Hubbard's Cave offers more than 6,000 feet of passage to explore, as well as the legendary "Mystery Pit," which entertained early visitors to this undeveloped cave.

passageways. Eventually, through a series of squeezes, the caver emerges at Jones Beach, the end of the dry cave. Beyond Jones Beach, wetsuits are required to continue through the flooded passages, which range in depth from a few inches to well over the top of your head.

Closer to Glenwood Springs, Hubbard's Cave is the most difficult of these three popular caves to reach, but it is the easiest to traverse. After a rough, steep, four-wheel-drive road from Lookout Mountain, the final half-mile to the cave consists of a sometimes-exposed foot trail following the limestone cliffs of Glenwood Canyon's southern rim.

Cavers should record their visit in the Forest Service register at the middle of the three entrances. From this register, a rock stairway descends into the cave. Hubbard's Cave has more than 6,000 feet of passage, much of it of comfortable walking dimensions. Four main parallel passages are connected by cross passages with high side alcoves reached by climbing dubious wooden ladders.

Off the East Entrance parallel, an excavated side stoopway proceeds to a blind pit. This is the legendary "Mystery Pit," where early visitors would drop rocks, never hearing the stones hit bottom. Though the pit allegedly plunged 1,400 feet deep (extending far below the limestone and even the floor of Glenwood Canyon—a geologic impossibility!), it later became "lost" when rocks blocked the shaft. Careful investigation of the pit in 1982 established that it actually connects to the cave's easternmost passage, a dirt-floored walkway that could accept plenty of tossed rocks without a sound.

Many enjoy negotiating the low crawls from the cave's West Entrance parallel to the Grape and Gypsum rooms. The Gypsum Room ranks as the cave's largest and most inspiring chamber, with high domes and extensive crystalline gypsum deposits.

Adjacent to Hubbard's Cave are two smaller caves, Column and Ice. The latter sometimes has ice, but is better known for its exceptionally high domes. The former boasts about 200 feet of passage, much of which involves crawling. It is notable for a joke played on cavers who visited in 1974, when someone sighted two glowing, unblinking eyes at the back of the cave. Fearful it was a mountain lion or a bear, the cavers cautiously backed away. That is, until the bravest soul decided to get a better look at the cornered animal. He discovered the "eyes" were actually pieces of reflective tape placed on the wall.

OPPOSITE: David Harris straddles the water in Thunder Road while setting up a photograph of Spring Cave's underground stream; INSET: a cluster of calcite botryoids in Hubbard's Cave harbors droplets of water.

A Primer on Underground Travel

If you're planning on venturing underground, you'll want to be certain you are adequately prepared. Caves are unforgiving places with plenty of real dangers. More than once in recent years, novice cavers have found themselves in life-threatening situations because of poor judgment or inadequate preparation. People can get lost, seriously injured, or even killed in caves.

For most beginning cavers, a commercial cave is the proper starting point. By touring a commercial cave with electric lighting, you can find out if you're claustrophobic and if exploring underground passageways is of further interest. Many commercial operations offer a variety of tours led by knowledgeable guides. By progressing to one of the more challenging tours, you'll discover if you're physically fit to continue your explorations.

Part of the preparation for going into caves is mental. Unlike the surface world, the sun never rises in caves beyond the limited twilight zone at the entrance. No matter how long you wait, your eyes never adjust to the total and complete darkness underground. Also remember that caves are natural features not designed for safe and easy travel. Even the easiest caves have low or tight passageways, unstable rocks, and unforeseen drop-offs. Many caves demand serious climbing and even ropework in precipitous shafts. Mud and slippery surfaces can cause trouble, as can small crawlways that might be too tight for traversing. Some caves even flood unexpectedly because of rising streams or heavy rain outside.

Caution is key in cave exploration. A common rule in any trip is that the group is only as strong as its weakest member. If a party member is unwilling to progress beyond a particular obstacle, the team turns around or chooses another destination. Team members always assist and support each other while underground.

Of course, no one should venture into a cave alone. Though teams might occasionally split up for a few minutes to examine possible leads, they always set a time and location to reconvene. Team members also alert others of their underground intentions. Registers at cave entrances serve as handy references if rescuers have to find the group in case of trouble. It's also wise to leave a note in your parked vehicle and communicate with a responsible friend or family member before you enter a cave. Be absolutely certain someone outside knows where your team is going and when you plan to return.

Detailed knowledge of the cave you're visiting can help you be certain you have the proper gear for the challenges that await you. Ask yourself these questions: What type of cave am I visiting? Is it warm? Cold? Wet? Dusty? Vertical? Horizontal? Do I have a map?

To visit any cave, underground travelers need some basic caving gear. In Colorado, most caves are cold, with temperatures ranging from the mid-30s to the low 50s. Warm clothing, preferably in layers, is a necessity. Hypothermia is a danger in Colorado caves for even experienced cavers. Clothing need not include down coats, but plan to get muddy or dusty. Many cavers choose coveralls, either heavy cotton garments or special nylon coveralls from

Even on a commercial cave tour, visitors such as these traveling the main passage of Manitou Cave wear warm clothing, work gloves, sturdy boots, and hardhats with headlamps—and plan on getting dirty.

sporting and caving shops. Wool sweaters and long underwear can also protect you from the cold.

Cavers concerned about the wear and tear on their knees also invest in a good pair of knee pads; wrestler-style knee pads worn under coveralls work best. Sometimes cavers also use elbow and wrist pads.

Cavers usually wear gloves while traveling underground. These range from inexpensive cotton gloves from the hardware store to fine leather gloves. Because you constantly use your hands underground, the gloves must be flexible and allow you to grip handholds.

Sturdy boots are essential in any cave. Sneakers and running shoes don't provide the traction needed to climb slippery rocks or walls. Leather climbing boots with neoprene soles are helpful, though caves and mud can take their toll on leather. Waterproofing your boots is a good preventative measure, particularly for touring caves known to have water or deep mud.

Even novices should wear helmets when visiting caves, as low ceilings, overhangs, and rock projections are all prime spots to bump your head. Inexpensive plastic construction helmets with elastic chinstraps and no rim are a popular choice, though serious cavers planning more than a single easy trip or intending anything in a vertical cave should purchase a well-built rock-climbing helmet with a three-point chinstrap. Helmets also protect your head from falling rocks knocked down by cavers above you.

During the last 20 years, lights for caving have progressed from acetylene lamps to sophisticated electric lights running on lithium, alkaline, or rechargeable batteries.

Armed with the appropriate caving gear for a safe journey, Bill Allen and Alan Williams take in the view at the Groaning Cave entrance.

Careful not to disturb the fragile underground environment, Christy Harrison perches above a pristine lake in Glenwood Caverns' Beginner's Luck.

Serious cavers always sport helmet-mounted lights, as caving often calls for the use of both hands.

Additional independent light sources must be part of any caving kit. Most cavers bring underground a small, side-mounted pack containing two or three extra lights. Secondary light sources range from an extra helmet-mounted light to a flashlight to chemical light sticks. Candles are commonly carried, not so much to light the way, but to use as a source of heat in case you have to wait for rescue in a cave.

Other materials carried in packs include a snack and water, a drop cloth for capturing crumbs from your underground meal, extra batteries for lights, a small first-aid kit, and even a large, plastic trash bag to sit in as a heat tent (the candle being the source of heat). Visitors using glasses or contact lenses might carry along a second set. An increasing number of cavers are also bringing bottles and other containers in which to remove all bodily waste from caves.

When exploring a cave, always look back to see where you have come from. Even if you're using maps, many passageways can be confusing. It isn't uncommon to momentarily become uncertain of your route. However, by carefully examining the passageways and calmly remembering what you have already seen and what obstacles you have passed, it's possible to find your way back out. String and arrows aren't needed for navigation.

Visiting caves can be an enjoyable and exciting adventure for people of all ages. With thoughtful preparation, your journeys into underground worlds can be accomplished safely and successfully.

Many other caves are found throughout Colorado. Some, like those in Williams Canyon near Manitou Springs, are privately owned and may not be explored without permission from the owners. Others, including the small caves at Rifle Falls State Park, are open for adequately prepared visitors. No special permission or permits are required to venture into the majority of the state's caves, though many of the larger caves have registers at their entrances that should be signed.

Individuals interested in serious exploration and study of Colorado caves might seek out membership in a local chapter of the National Speleological Society (see Appendix B). Colorado has several active chapters that meet on a regular basis. Newcomers are welcome at the meetings, during which future trips and projects are arranged, previous trips are reported on, and fellowship abounds. *Rocky Mountain Caving*, a quarterly journal, publishes news on caving in Colorado and the Rocky Mountain region.

Like all Williams Canyon caves, Three Hole Cave is privately owned—only cavers with permission and an authorized guide may enter.

Afterword

"I knew that I had made a great discovery, and as I lay there, with my head and shoulders out of the hole, holding my candle above my head to light up the room in all directions as far as possible, the sight was so deeply imbedded in my memory that it can never be effaced. It was as though Aladdin with his wonderful lamp had effected the magic result."
—*George W. Snider, on his 1881 discovery of Cave of the Winds' Canopy Hall in his 1915 book,* How I Found and How I Lost the Cave of the Winds and the Manitou Grand Caverns

More than a century has passed since George W. Snider, a 30-year-old stonecutter from Ohio, unearthed this most important finding at Cave of the Winds. History would be far different today if Snider had not followed the breeze that nearly extinguished his candle deep within the abandoned commercial cavern. While he had visions of the grandeur and wealth his discovery would bestow upon him, he could never have foreseen the growth of not only Cave of the Winds, but also Colorado caving throughout the next century and beyond.

While relatively few ventured underground for pleasure or knowledge in Snider's time, by the end of the 20th century visitors to Colorado's caverns numbered in the hundreds of thousands. Though the commercial caves attract a great majority of these visitors, cavers of all abilities increasingly seek to explore Colorado's undeveloped caves.

In 1951, Dr. William R. Halliday, founder of the Colorado Grotto, had difficulty gathering together the minimum number of cavers needed to charter a National Speleological Society chapter. Today, an estimated 1,000 cavers make their homes in Colorado, most living in the major metropolitan areas of the Front Range. With only one-fourth affiliated with formal caving groups, cooperation, communication, and cave protection grow progressively more difficult.

These delicate worlds are currently receiving greater impact than at any time in their long history. More and more cavers crawling and squeezing and pushing each and every passage bring about changes that might take millennia to reverse if left to natural processes. Recognizing this continuing harm, many cavers turn to secrecy to protect unique and pristine discoveries. Some depend on escalating protection and regulation from private and public landowners and managers; in some states, cave protection laws assessing stiff penalties for damage offer assistance. Others seek education as a means of safeguarding caves, attempting to teach other cavers and the public about the

ABOVE: **By the 1950s, when tours still cost only a dollar, Cave of the Winds had a paved parking lot for the era's grandiose automobiles.** PHOTO COURTESY CAVE OF THE WINDS COLLECTION.
OPPOSITE: Spring Cave, a popular "wild" cave managed by the White River National Forest, is open to the public for careful exploration.

uniqueness and exceptional fragility of these darkened worlds and their life.

Although rising numbers of visitors bring degradation to caves, more cavers also open the possibility of additional discoveries. Energetic cavers have found remarkable new caves in recent years and will continue to make fresh revelations. Caves unimagined today will be dug open in hidden, obscure locations. Curious explorers will also expand currently known caves.

Each generation proclaims its own era the "golden age" of Colorado caving. Perhaps not surprisingly, the next generation usually makes even more discoveries. Knowledgeable geologists suggest that although there is a finite number of caves within this state, only a small percentage might be known so far.

Like the great wilderness explorers of our past, today's cavers push the frontiers of the underground and come upon extraordinary new lands. Discovery remains a prime motivator —opening previously unknown caves and passageways and identifying new life-forms and natural processes. Science and technology play as much of a role in caving as hard hats and lights, for these bodies of knowledge heighten the opportunity for discovery.

All of Colorado's caving history—from Arthur Love's tentative 1869 exploration of Cave of the Winds to recent explorations by Donald G. Davis, Mike Frazier, Evan Anderson, Paul Burger, Doug and Hazel Medville, Gene Dover, and many others—may be but a prelude to secrets as yet locked within the vast and mighty Rocky Mountains. Intelligent, systematic investigations will help unveil many of these hidden splendors.

Just as we face the consequences of decisions made by those who passed before us, those who follow us depend on the choices we make. In actively protecting our magical underground worlds, we help to preserve our own heritage. Let us establish a legacy of responsible cave stewardship— with the hope that future generations will recognize our wisdom and follow closely in our footsteps.

᭸

L. P. Lawrence savors her caving trip to Breezeway Cave's Celestial City.

NOTE: Because tour routes, hours, and seasons may change, please contact the caves for the most current information.

Cave of the Winds

P.O. Box 826
Manitou Springs, CO 80829
719.685.5444
www.caveofthewinds.com

In Cave of the Winds' Valley of Dreams, Megan Gray and Joseph Blackshaw pass by the Pickett Tablet, one of a handful of "cave shield" formations in Colorado.

Directions: Follow Highway 24 to Manitou Springs six miles west of Colorado Springs (take Exit 141 off Interstate 25). The cave is reached via an all-season, paved highway overlooking Manitou Springs. Serpentine Drive is steep and switchbacking, so use low gear both uphill and downhill and stay to the right.

Season: Open year-round, reduced hours in winter months.

Other Attractions: Laser Canyon light and music show Memorial Day through Labor Day; self-guided nature trail to undeveloped Snider's Cave; views of Williams Canyon.

Visitor Tours:

Discovery Tour: Visitors follow an electrically lit tour route through the Cave of the Winds. The paved tour route with numerous stairs features Canopy Hall, Temple of Silence, Valley of Dreams, and Bridal Chamber.

Manitou Grand Caverns Lantern Tour: Costumed guides bring visitors back to the turn of the 20th century on a historic tour through Manitou Grand Caverns. Visitors carry kerosene lanterns to illuminate the passageways, including the Grand Concert Hall, Lover's Lane, Grant's Monument, and the Fairy Bridal Chamber. The route includes gravel pathways and a few stairs.

Manitou Cave Explorer's Tour: A physically challenging trip into the formerly commercial Manitou Cave. Visitors must be able to climb and crawl through often-narrow cave passageways, and should wear old warm clothes, sturdy hiking boots, and gloves. The route can include Whirlpool Dome and the Centipede and Deepwater sections. Tours are scheduled on a daily basis during the summer, and on weekends in the winter (reservations are required).

Glenwood Caverns

508 Pine Street
Glenwood Springs, CO 81601
970.945.4CAV
800.530.1635
www.glenwoodcaverns.com

Glenwood Caverns developer Steve Beckley pauses in King's Row as lighting is installed along the tour route.

Directions: The visitor center and museum is located on Pine Street behind the historic Hotel Colorado in Glenwood Springs. Take Exit 116 off Interstate 70, turn east on Highway 82, and follow it to Pine Street. Free shuttles ferry visitors up and down Iron Mountain. (A proposal to build a tram linking the caves with downtown Glenwood Springs—which would allow the caves to remain open year-round—is pending approval.)

Season: Open April through October.

Other Attractions: Views of Glenwood Canyon.

Visitor Tours:

Glenwood Caverns Family Tour: Visitors enjoy an electrically lit jaunt through the historic Fairy Cave. The gravel tour route features several underground chambers and the spectacular view of Glenwood Canyon from Exclamation Point.

Guides then lead visitors on an excursion into the lower portion of Glenwood Caverns. The electrically lit route features the fantastically decorated King's Row and The Barn. The route includes gravel pathways and numerous stairs.

Wild Tour: A physically challenging tour off the commercial trail in Glenwood Caverns. Visitors must be able to climb and crawl through often-narrow cave passageways, and should wear old warm clothes, sturdy hiking boots, and gloves. The route can include the Pendant Room, the Drum Room, and other rooms. Tours are scheduled on a daily basis during the summer (reservations are required).

Yampah Spa Vapor Caves

709 E. 6th Street
Glenwood Springs, CO 81601
970.945.0667
www.glenscape.com/spas.htm
www.visitvailvalley.com/spas/yampah.htm

Directions: Yampah Spa Vapor Caves is located just east of the Hot Springs Pool in Glenwood Springs. Take Exit 116 off Interstate 70, then turn east on Highway 82 (6th Street).

Season: Open year-round.

Other Attractions: Massage therapy, hot tubs, and other health and beauty treatments.

Visitor Tours:

Vapor Caves: Visitors don't tour the Vapor Caves, the only known natural vapor caves in North America, but sit on benches in the geothermal underground steam bath.

Appendix B
Caving Organizations and Websites

Hazel Barton stands next to a stalagmite tipped with aragonite crystals in Glenwood Caverns' Paradise.

National Caving Organizations

American Cave Conservation Association

P.O. Box 409, Horse Cave, KY 42749; phone: 270-786-1466; website: www.cavern.org. National membership association for those interested in cave conservation and protection. Publishes a quarterly magazine, *American Caves*, and manages American Cave Museum and Hidden River Cave in Horse Cave, Kentucky.

Bat Conservation International

P.O. Box 162603, Austin, TX 78716; phone: 512-327-9721; website: www.batcon.org. An international nonprofit educational association dedicated to the study and protection of bats (not a caving organization). Publishes quarterly *BATS* magazine.

Cave Research Foundation

Website: www.cave-research.org. Scientific-oriented nonprofit organization with cave and karst exploration, survey, and science programs in several states, including Kentucky, New Mexico, and California. Publishes quarterly *CRF Newsletter*.

National Caves Association

4138 Dark Hollow Rd., McMinnville, TN 37110; phone: 931-668-3925; website: www.cavern.com. National commercial caves trade association with more than 90 members in the United States and Bermuda.

National Speleological Society

2813 Cave Ave., Huntsville, AL 35810; phone: 256-852-1300; website: www.caves.org. National membership association for cave exploration, conservation, and science. Publishes monthly *NSS News* magazine. Has more than 12,000 members and 200 local chapters nationwide, including five in Colorado:

- Colorado Grotto (Denver)
- Front Range Grotto (Northglenn)
- Northern Colorado Grotto (Fort Collins/Loveland)
- Southern Colorado Mountain Grotto (Colorado Springs)
- Timberline Grotto (Glenwood Springs)

Caving Websites: References and Links

http://web.infoave.net/~tcurtis/links.htm
Helpful cave and caving links page.

www.amazingcaves.com
Official website for the large-format film
Journey Into Amazing Caves.

www.aqd.nps.gov/grd/geology/caves/index.htm
National Park Service Cave and Karst Resources website;
includes informative *Inside Earth* newsletter.

www.ednet.lancs.ac.uk/luss/www/index.html
Large international cave and caving links page.

www.goodearthgraphics.com/virtcave.html
The educational "Virtual Cave," featuring photos and
information about caves.

www.gorp.com/gorp/activity/caving.htm
Great Outdoor Recreation Pages website features good links to
cave, caving club, and individual web pages.

www.mysteries-megasite.com/frame.html
Huge Internet cave and caving links page; claims minimum of
1,000 links.

Caving Websites: Retail

http://mgmtsys.com
Karst Sports, a West Virginia–based caving supplies company.

www.bluewater-climbing.com
Georgia-based manufacturer of caving and climbing rope.

www.petzl.com
French-manufactured Petzl headlamps and other caving
equipment, available worldwide.

www.pmirope.com
Pigeon Mountain Industries, a Georgia-based manufacturer of
caving rope.

www.speleobooks.com
Speleobooks, one of America's largest cave book vendors;
proprietor Emily Davis Mobley is a former Colorado Grotto
member.

www.speleo.co.uk
Speleo Technics, a headlamp manufacturer that exports lamps
worldwide from the United Kingdom.

www.speleoprojects.com
Swiss-based cave book publisher Speleo Projects hosts this
multilingual website; if you see a book you like, have your
currency changer ready as prices are not in dollars!

Appendix C
Related Reading and Videos

Books

Borden, James D., and Roger W. Brucker. *Beyond Mammoth Cave: A Tale of Obsession in the World's Longest Cave.* Carbondale, Ill.: Southern Illinois University Press, 2000.
> *The continuing saga of the exploration of Mammoth Cave, the world's longest cave, including the connection to Roppel Cave.*

Brucker, Roger W., and Richard A. Watson. *The Longest Cave.* Carbondale, Ill.: Southern Illinois University Press, 1987.
> *An account of the exploration of Kentucky's Mammoth Cave, the longest known cave in the world.*

Burger, Paul, Lawrence Fish, Patricia Kambesis, and Steve Reames. *Deep Secrets.* St. Louis, Mo.: Cave Books, 1999.
> *Colorado cavers and the exploration of New Mexico's Lechuguilla Cave.*

Conn, Herb, and Jan Conn. *The Jewel Cave Adventure.* St. Louis, Mo.: Cave Books, 1981.
> *Documents the exciting exploration of South Dakota's Jewel Cave.*

Courbon, Paul, Claude Chabert, Peter Bosted, and Karen Lindsley. *Atlas: Great Caves of the World.* St. Louis, Mo.: Cave Books, 1989.
> *Descriptions and maps of some of the longest and most interesting caves worldwide.*

Dasher, George. *On Station.* Huntsville, Ala.: National Speleological Society, 1994.
> *Detailed book on cave surveying techniques.*

Farr, Martyn. *The Darkness Beckons.* St. Louis, Mo.: Cave Books, 1991.
> *In-depth history of cave diving across the globe.*

Halliday, William R. *American Caves and Caving.* New York, N.Y.: Harper & Row, 1974.
> *Classic tales of American caves and caving.*

———. *Depths of the Earth.* New York, N.Y.: Harper & Row, 1976.
> *More classic stories of American caves and caving.*

Hill, Carol, and Paolo Forti. *Cave Minerals of the World.* Huntsville, Ala.: National Speleological Society, 1997.
> *Comprehensive book on cave mineralogy.*

Howes, Chris. *To Photograph Darkness.* Carbondale, Great Britain: Gloucester/SUI Press, 1989.
> *History of cave photography.*

McClurg, David. *Adventure of Caving.* Vallejo, Calif.: D & J Press, 1996.
> *Beginner's guide to cave exploration.*

Murray, Robert K., and Roger W. Brucker. *Trapped!* Lexington, Ky.: University Press of Kentucky, 1982.
> *Well-researched documentation of the attempts to rescue caver Floyd Collins from Kentucky's Sand Cave in 1925.*

Nelson, Jim. *Glenwood Caverns and the Historic Fairy Caves.* Glenwood Springs, Colo.: Blue Chicken Inc., 2000.
> *Colorful book about Colorado's Glenwood Caverns featuring photographs by David Harris.*

Nymeyer, Robert. *Carlsbad, Caves and a Camera*. St. Louis, Mo.: Cave Books, 1978.
>   *Exploration of caves in and around Carlsbad Caverns National Park, New Mexico, during the 1930s.*

Rea, G. Thomas, ed. *Caving Basics*. Huntsville, Ala.: National Speleological Society, 1992.
>   *Comprehensive introduction to caving techniques.*

Rhinehart, Richard. *Without Rival: The Story of the Wonderful Cave of the Winds*. Virginia Beach, Va.: The Donning Company Publishers, 2000.
>   *History and geology of Colorado's famous Cave of the Winds, with photographs by David Harris.*

Schultz, Ron. *Looking Inside Caves and Caverns*. Santa Fe, N.M.: John Muir Publications, 1993.
>   *Children's book about caves.*

Smith, Bruce, and Allen Padgett. *On Rope*. Huntsville, Ala.: National Speleological Society, 1996.
>   *Excellent guide to rope techniques for vertical climbing and rappelling in caves.*

Taylor, Michael Ray. *Dark Life*. New York, N.Y.: Scribner, 1999.
>   *Entertaining book on the science behind microbiological life in caves; includes Colorado cavers and events.*

———, ed. *Lechuguilla: Jewel of the Underground*. Basel, Switzerland: Speleo Projects, 1998.
>   *Beautifully illustrated book about New Mexico's famed Lechuguilla Cave.*

Thompson, Norman R., and John van Swearingen. *On Caves and Cameras*. Huntsville, Ala.: National Speleological Society, 2001.
>   *Complete guide to photography in caves.*

Tuttle, Merlin. *America's Neighborhood Bats*. Austin, Texas: University of Texas Press, 1997.
>   *Informative book about bats in the United States.*

Videos

Bats—Predators of the Wild. Burbank, Calif.: Warner Home Video, 1994. *Excellent documentary about bats.*

Lechuguilla Cave: The Hidden Giant. Denver, Colo.: Denver Museum of Nature & Science, 1988.
>   *Exploration of New Mexico's Lechuguilla Cave.*

Mysteries Underground. Washington, D.C.: National Geographic Society, 1992. *Television documentary featuring exploration and science in caves throughout the United States, including New Mexico's Lechuguilla Cave.*

Silent Splendor. Denver, Colo.: Denver Museum of Nature & Science, 1986.
>   *Discovery and scientific study of Cave of the Winds' spectacular Silent Splendor passage.*

Spirit of Exploration. Carlsbad, N.M.: Carlsbad Caverns/Guadalupe Mountains Association, 1993.
>   *The caves of Carlsbad Caverns National Park, including Carlsbad, Lechuguilla, Slaughter Canyon, Ogle, and Spider.*

Strange Creatures of the Night. Washington, D.C.: National Geographic Society, 1995.
>   *Television documentary on life that exists in darkness, including bats and other cave creatures.*

Wind Cave: The World Below. Huntsville, Ala.: National Speleological Society, 1991.
>   *Geology, history, and exploration of South Dakota's Wind Cave.*

Index

Note: Citations followed by the letter "p" denote photos.

A

Alexander's Cave. *See* Cave of the Clouds
American Cave Conservation Association, 128
Aragonite
    definition of, 18
    photograph of, 81p
Ascenders, definition of, 18

B

Bair Cave, 20p, 41
Barton, Hazel, 94, 94p
Bat Conservation International, 128
Beginner's Luck (Glenwood Caverns), 22p, 32p, 102p, 121p
Blowholes, definition of, 18
Blue Pool Room (Groaning Cave), 49p
Bones, in caves, 66–67, 66p
Botryoids
    definition of, 18
    photographs of, 18p, 46p, 119p
Braided Cave, 53p
Breakdown
    definition of, 18
    photograph of, 14p
Breezeway Cave
    Celestial City, 124p–125p
    conservation efforts, 100, 103
    Elkhorn Chambers, 19p, 64p, 69p, 111p
    exploration of, 25, 28
    Hangman's Hole, 26p–27p
    Happy Trails, 60p
    Heaven's Gate, 2p
    Holy Waters, 19p, 102p
    photographs of, 9p, 17p, 101p
    Stone River, 8p, 59p, 90p
Bridal Chamber (Cave of the Winds), 106p, 108p, 114p, 118p
Buffalo Cave, 47

C

Calcite
    definition of, 18
    photographs of, 9p, 81p
Canopy Avenue (Manitou Grand Caverns), 89p
Canopy Hall (Cave of the Winds), 88p, 97p
Canyon Creek Cave, 41
Carbonic acid, definition of, 18
Cathedral Spires (Cave of the Winds), 106p, 108p

Cave bacon
    definition of, 18
    photographs of, 18p, 95p
Cave conservation. *See* Conservation, cave
Cave coral, definition of, 18
Cave Creek Cavern
    Colorado Room, 15p
    discovery of, 42
    reopening of, 9, 11, 12, 14–15
Cave formation. *See* Formation, cave
Cave hunting, definition of, 18
Cavelets, definition of, 18
Cave life, 60–61, 64
Cave names, 30
Cave of the Chimes, 106p
Cave of the Clouds
    destruction of, 12, 15
    development of, 86
    history of, 38
    photographs of, 13p, 38p, 87p
Cave of the Fairies (Fairy Cave)
    development of, 86
    Exclamation Point, 16p
    exploration of, 44
    history of, 38–39
    photographs of, 16p, 86p
    See also Glenwood Caverns
Cave of the Logs, 40
Cave of the Winds
    advertising pamphlet for, 16p
    Bridal Chamber, 106p, 108p, 114p, 118p
    Canopy Hall, 88p, 97p
    Cathedral Spires, 106p, 108p
    cleaning of, 97, 99
    conservation efforts, 100, 103
    development of, 85–86, 88–89, 91
    directions for reaching, 126
    early photographs of, 108
    entrance building, 16p, 39p
    exploration of, 22, 25
    formation of, 57–58
    history of, 37–38, 39
    map of, 73p
    Mirror Lake, 99p
    Native Americans and, 40
    Oriental Garden, 98p, 99p
    photographs of, 35p, 40p, 74p, 82p, 85p, 103p, 123p
    Reception Hall, 3p

Cave of the Winds *continued*
   Silent Splendor, 24p, 25p, 65p
   surveying of, 73–74
   Temple of Silence, 18p, 30p, 109p
   tours of, 114, 117, 126
   Valley of Dreams, 126p
Cave pearls, definition of, 18
Cave regions, Colorado, 55p
Cave Research Foundation, 128
Cave restoration. *See* Restoration, cave
Caverns, definition of, 18
Cavers, definition of, 18
Caves, definition of, 18
Caves of Colorado, 7
Cave surveying. *See* Surveying, cave
Celestial City (Breezeway Cave), 124p–125p
Centipede Cave, 32
Chimneys, definition of, 18
Cliffhanger Cave, 73–74, 96p
Colorado Grotto, 11, 42, 44
Colorado Room (Cave Creek Cavern), 15p
Column Cave, 118
Columns
   definition of, 18
   photographs of, 6p, 116p
Commercial caves, definition of, 18
Conservation, cave
   current practices, 100, 102, 103
   importance of, 7, 15, 123, 125
   respecting property rights, 103
CSU Passage (Groaning Cave), 23p, 47p, 48p

D
Dangers of caving, 32, 120
Darkness, cave, 12
Devil's Den, 42
Digs, definition of, 18
Dogtooth spars, definition of, 18
Dolomite, definition of, 18
Domes, definition of, 18
Donahue's Cave, 76–77
Draperies
   definition of, 18
   photographs of, 8p, 14p, 47p, 84p, 95p, 115p
Dripstones, definition of, 18

E
Easter Dome, 76
Elkhorn Chambers (Breezeway Cave), 19p, 64p, 69p, 111p
Emerald Pool (Spring Cave), 31p
Exclamation Point (Cave of the Fairies), 16p

F
Fixin'-to-Die Cave, 47
Flowstone
   definition of, 19
   photographs of, 19p, 87p, 97p
Formation, cave
   fracture caves, 53
   glacial outwash process, 60, 62
   granite caves, 62
   gypsum caves, 62
   hot springs caves, 55
   hydrogen sulfide caves, 53, 55
   hydrogen sulfide process, 60
   illustration of, 54p
   mixing waters process, 57–58, 60
   mud caves, 62
   snowfield ice caves, 53
   solutional caves, 53
   tufa caves, 53
Fulford Cave
   accidents in, 32
   discovery of, 41–42
   exploration of, 44
   photographs of, 52p, 63p, 112p, 113p, 116p
   Two-Level Room, 10p
   visiting, 117

G
Gear, caving, 120–121
Glenwood Caverns
   Beginner's Luck, 22p, 32p, 102p, 121p
   development of, 83, 85, 92–93
   directions for reaching, 127
   exploration of, 21–22, 29, 32
   formation of, 58, 60
   Jam Crack, 29p
   King's Row, 80p, 84p, 91p, 127p
   Lower King's Row, 58p, 105p
   Paradise, 6p, 18p, 110p, 128p
   photographs of, 14p, 17p, 21p, 44p, 78p–79p, 83p, 92p–93p, 95p, 115p
   surveying of, 75
   tours of, 114, 117, 127
   See also Cave of the Fairies (Fairy Cave)
Glossary of terms, 18–19
Goose Creek Cave, 33
Grand Concert Hall (Manitou Grand Caverns), 57p
Grape Room (Hubbard's Cave), 18p
Groaning Cave
   Blue Pool Room, 49p
   conservation efforts, 100
   CSU Passage, 23p, 47p, 48p
   discovery of, 47

Groaning Cave *continued*
    exploration of, 22
    photographs of, 36p, 46p, 47p, 71p, 81p, 120p
    surveying of, 71–72, 75, 77
Grottos, definition of, 19
Guano, definition of, 19
Gypsum, definition of, 19
Gypsum flowers
    definition of, 19
    photograph of, 46p

H
Hangman's Cave, 41
Hangman's Hole (Breezeway Cave), 26p–27p
Happy Trails (Breezeway Cave), 60p
Heaven's Gate (Breezeway Cave), 2p
Helictites
    definition of, 19
    photographs of, 2p, 17p, 19p, 24p, 59p, 64p, 65p, 69p, 111p
History, Colorado caving, 16–17
Holy Waters (Breezeway Cave), 19p, 102p
Horseshoe Tunnel (Manitou Grand Caverns), 108p
Hourglass Cave, 32, 67
Hubbard's Cave
    discovery of, 42
    Grape Room, 18p
    Mystery Pit, 118p
    photographs of, 1p, 43p, 44p, 58p, 70p
    visiting, 118
Humidity, cave, 12

I
Ice Cave, 118
Indian Cave, 104p

J
Jam Crack (Glenwood Caverns), 29p
Joints, definition of, 19
Journey Into Amazing Caves, 94

K
Karst, definition of, 19
Kimball Cave, 41
King's Row (Glenwood Caverns), 80p, 84p, 91p, 127p

L
Lava tubes, 53, 63–64
Leads, definition of, 19
Leavenworth (Narrows Cave), 17p, 34p
Life, in caves, 60–61, 64
Limestone, definition of, 19
Lower King's Row (Glenwood Caverns), 58p, 105p

M
Mammoth Cave
    development of, 85
    history of, 37
    surveying of, 73
Manitou Cave
    development of, 86
    entrance building, 37p
    history of, 37, 48
    lost tour of, 50–51
    photograph of, 120p
    tours of, 117, 126
    Whirlpool Dome, 50p
Manitou Grand Caverns
    Canopy Avenue, 89p
    development of, 86, 88–89, 91–92
    early photographs of, 108
    exploration of, 22, 25
    Grand Concert Hall, 57p
    history of, 37–38
    Horseshoe Tunnel, 108p
    map of, 74p
    photographs of, 38p, 85p, 89p
    surveying of, 73
    tours of, 117, 126
Mesa Verde National Park, 40, 40p
Mining, and cave discoveries, 39, 41–42
Mirror Lake (Cave of the Winds), 99p
Moonmilk
    definition of, 19
    photograph of, 6p
Moonmilk Cave, 41
Mystery Pit (Hubbard's Cave), 118p

N
Naming of caves, 30
Narrows Cave
    exploration of, 34–35
    photographs of, 11p, 58p, 66p
    Leavenworth, 17p, 34p
    Zephyr Avenue, 35p
National Caves Association, 128
National Speleological Society, 121, 128
Native Americans, and caves, 40
New Cave. *See* Manitou Cave

O
Octagon Cave, 76p
Organizations, caving, 128
Oriental Garden (Cave of the Winds), 98p, 99p

P
Paradise (Glenwood Caverns), 6p, 18p, 110p, 128p
Parris, Lloyd, 7

Pedro's Cave. *See* Manitou Cave
Photographing caves
    early attempts at, 108
    recommendations for, 109
    understanding light, 105–107, 108, 110
Pits, definition of, 19
Porcupine Cave
    bones in, 67
    discovery of, 42
    photographs of, 42p, 67p
Premonition Cave, 41, 47

R
Radford Cave, 42
Rappel, definition of, 19
Reception Hall (Cave of the Winds), 3p
Regions, Colorado cave, 55p
Restoration, cave, 99–100, 100p
Rimstone
    definition of, 19
    photographs of, 9p, 46p
Rinehart, Charles, 88–89
Rocky Mountain Caving, 121

S
Safety, caving, 32, 120
Shamrock Cave, 41
Silent Splendor (Cave of the Winds), 24p, 25p, 65p
Sinkholes, definition of, 19
Skeleton Cave, 41
Snider, George W., 85p, 88–89, 88p
Snider's Cave, 62p
Solution, definition of, 19
Spanish Cave
    discovery of, 41
    exploration of, 44, 45
    photographs of, 44p, 45p
Spectre Cave, 47
Speleology, definition of, 19
Speleothems, definition of, 19
Spring Cave
    discovery of, 41
    Emerald Pool, 31p
    exploration of, 28–29, 31, 44, 47
    map of, 75p
    photographs of, 19p, 29p, 31p, 60p, 122p
    Thunder Road, 119p
    visiting, 117–118
Stalactites
    definition of, 19
    photographs of, 6p, 18p, 19p, 25p, 81p, 90p, 99p, 116p
Stalagmites
    definition of, 19
    photographs of, 18p, 25p, 46p, 84p, 128p

Stone River (Breezeway Cave), 8p, 59p, 90p
Summer's End Cave, 29, 47
Surveying, cave
    challenges of, 72–73
    early surveys, 73
    impact of computer technology on, 72
    "survey-as-you-go" doctrine, 22
Sweetwater Indian Cave, 40

T
Temperature, cave, 12
Temple of Isis, 107p
Temple of Silence (Cave of the Winds), 18p, 30p, 109p
Tesla Coil, 69p
Three Hole Cave, 121p
Thunder Road (Spring Cave), 119p
Thursday Morning Cave, 47
Timeline, Colorado caving, 16–17
Twenty Pound Tick Cave, 17p, 47
Two-Level Room (Fulford Cave), 10p

U
Unger, Donald Bruce, 33, 33p
Ute Indian Pictograph Cave, 40
Ute Mountain Tribal Park, 40

V
Valley of Dreams (Cave of the Winds), 126p
Vapor Cave, 40
Virgin caves, definition of, 19

W
Water tables, definition of, 19
Websites, caving, 129
Whirlpool Dome (Manitou Cave), 50p
Whistling Cave, 41
White Marble Halls, 41
White River Plateau, 28p
Wild caves, definition of, 19
Williams Canyon, 5p, 16p, 51p, 56p, 76p, 107p
Wilson's Cave, 7p

Y
Yampah Spa Vapor Caves
    directions for reaching, 127
    history of, 38
    tours of, 127

Z
Zephyr Avenue (Narrows Cave), 35p